SIGNAL PROCESSING, MODULATION AND NOISE

Science and Technology Series

General Editor
PROFESSOR J. H. CALDERWOOD
M. Eng., Ph.D., C.Eng., F.I.E.E., F.Inst.P.
Chairman of Electrical Engineering Department
University of Salford

A GUIDE TO ADVANCED ELECTRICAL ENGINEERING
R. V. Buckley, B.Sc., Ph.D., C.Eng., M.I.E.E.

EXAMPLES IN ADVANCED ELECTRICAL ENGINEERING
J. F. W. Bell, B.Sc., Ph.D., C.Eng., M.I.E.E.

THEORY AND PRACTICE OF DRAWING FOR ENGINEERS IN S.I. UNITS
A. W. Barnes, C.Eng., M.I.Mech.E.,
and
A. W. Tilbrook, C.Eng., M.I.Mech.E., F.B.H.I.

INTRODUCTION TO ADVANCED ELECTRICAL ENGINEERING
C. Jones, B.Sc., C.Eng., M.I.E.E., A.M.Brit.I.R.E.

PRINCIPLES OF AUTOMATIC CONTROL
Martin Healey, M.Sc., C.Eng., M.I.E.E.

COMPLEX PERMITTIVITY
B. K. P. Scaife, B.Sc., Ph.D., M.A., F.I.P., M.I.E.E.

Other Books of Interest

THE ART OF SIMULATION
K. D. Tocher, D.Sc.

INTRODUCTION TO THE MATHEMATICS OF SERVOMECHANISMS
J. L. Douce, M.Sc., Ph.D., C.Eng., M.I.E.E.

HIGH FREQUENCY COMMUNICATIONS
J. A. Betts, B.Sc., Ph.D., C.Eng., M.I.E.E.

LOW NOISE ELECTRONICS
W. P. Jolly, B.Sc., C.Eng., F.I.E.E.

THE TRANSMISSION AND DISTRIBUTION OF ELECTRICAL ENERGY
H. Cotton, M.B.E., B.Sc., C.Eng., F.I.E.E.
and
H. Barber, B.Sc.

SIGNAL PROCESSING, MODULATION AND NOISE

J. A. BETTS
Department of Electronics, University of Southampton

AMERICAN ELSEVIER PUBLISHING COMPANY, INC.

NEW YORK 1971

Published in the United States by
AMERICAN ELSEVIER PUBLISHING COMPANY, INC.
52 Vanderbilt Avenue, New York, N.Y. 10017

First printed 1971

Copyright © 1970 J. A. Betts

Library of Congress Catalog Card Number 75-130357
International Standard Book Number 0-444-19671-4

Printed and bound in Great Britain

GENERAL EDITOR'S FOREWORD

by

PROFESSOR J. H. CALDERWOOD

M.Eng., Ph.D., C.Eng., F.I.E.E., F.Inst.P.

Chairman of Electrical Engineering Department
University of Salford

We are living in a world in which communications have reached such a degree of sophistication that it causes us little surprise when the view as seen from the windows of a space-ship circling the moon is projected on a screen in our living rooms so that we only have to wait a second or so longer to see the same scene as that viewed by the men in space-ship itself. The study of the fundamental concepts which have been developed into engineering reality in order to make such feats possible is a fascinating one. An enormous gap may seem to exist between modern intercontinental communications systems, and the primitive tribesman sending his messages by interrupting the column of smoke rising from his fire by means of the judicious manipulation of his blanket. Yet there is a greater similarity than might be at first supposed: the operators of both systems, and not merely one of them, are depending heavily on the fundamentals of thermodynamics. If that causes any surprise, it should be removed on reading this book. Not only should surprise be removed, but knowledge gained about the exciting task of building communications systems on which the modern world depends, the fundamental ideas on which they are based, and the difficulties, some of which are themselves of a deeply fundamental kind, which are encountered and have to be surmounted.

PREFACE

The majority of recently published text-books on Signal Processing and Noise have been written for post-graduate study and the professional communications engineer. This book, however, with the exception of Chapter 7, is intended for the undergraduate student meeting Telecommunications for the first time.

The book is divided into two parts: the first, comprising eight chapters dealing with analogue and digital modulation and noise, is presented with the minimum number of 'pure' mathematical proofs in order that the student may gain an immediate appreciation of the communications problem without getting lost in the equally important, but nevertheless theoretically biased, aspects of Fourier Analysis. The latter, together with other topics such as the Convolution Integral, Introductory Statistical Communication Theory and Spectral Analysis, is dealt with extensively in the second part, which consists of three Appendices. It must be stressed that these appendices form an integral part of the book and cannot be ignored if a full understanding of Chapters 1 to 8 is desired. The student is recommended to proceed through the book in the order in which it is set out, consulting the Appendices whenever it becomes necessary—this is clearly indicated in the main text.

Whilst the first part of Chapter 7, dealing with pulse code modulation, is considered suitable for undergraduate training, the second, on delta modulation, is of post-graduate standard. It is included in this book for the sake of completeness, and is probably the first thorough treatment of delta modulation to be published in text-book form.

Finally, I should like to thank Mr. D. G. Appleby for his useful comments and advice on various parts of the book, and also all those of my students who have pointed out errors.

CONTENTS

1
INTRODUCTION

Have you read the Preface . . .? Yes? Then the purpose of the book is understood and we may proceed.

A message or piece of information invariably occurs in non-electric form and is the result of some physiological or physical phenomenon. The multifarious detail associated with a particular source can be considered as potential information, since only a fraction of it may constitute the relevant (or required) message. Take, for example, eyelid movements; it may be necessary to transmit all the details of movement or to convey only information of blink rate. In the latter case a humanly controlled selection process is arranged to extract solely the information of blinking.

Whatever the message may be, it is necessary to generate a corresponding electrical signal for the purpose of telecommunication. This is achieved by using a suitable transducer, for example the microphone for speech, the television camera for moving pictures and so on. These signals can be classified as either analogue or digital; the distinction is best illustrated by the following two examples.

The transducer output corresponding to speech will vary continuously over some arbitrary but finite amplitude range depending upon the loudness of the speech and the gain characteristics of the microphone. The instantaneous amplitude can have any value within this range and for this reason the signal is classified as analogue. On the other hand, the electrical signal corresponding to information presented as a complex pattern of holes or no holes in a paper tape can have only two amplitude levels, one representing a hole and the other the absence of a hole. This is known as a digital signal and when the number of levels is two, as in this example, it is in binary form. In general, however, the amplitude range can be quantized into a number of discrete levels and at any instant in time the signal can only be at one of these levels—this is referred to as an m-ary signal (see Fig. 1–1).

Before the transducer outputs can be transmitted via a line or radio circuit it is necessary to process the signal in a way that is best suited to the propagation conditions likely to be encountered. The aim of this

[1]

book is to establish the basic principles of signal processing and to determine the effect of interference on the received signal. Fig. 1–2 illustrates how these aspects fit in to the overall system of message origination and transmission.

Firstly, however, it is essential to define the information sources commonly encountered and also to establish their electrical characteristics. For our requirements the following classification will suffice:

Type of signal	Classification of transducer output
speech (telephony)	analogue
teleprinter (telegraphy)	digital
computer information (data) }	
still pictures (facsimile)	analogue
moving pictures (television)	analogue
physical parameters (telemetry)	analogue or digital
navigational information (radar)	analogue or digital
remote control (telecontrol)	analogue or digital

Telephony, telegraphy, facsimile and television are relatively straightforward to define and we shall use these as examples to develop the general principles of signal bandwidth.

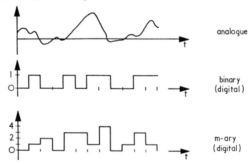

Fig. 1–1 Typical analogue and digital signals.

1–1 The Concept of Baseband

1–1.1 *Telephony*

The electrical output from a microphone varies in sympathy with the compressions and rarefactions of the sound wave and for speech can occupy a spectral range from 100 Hz up to the very high audio frequencies. A voice of low pitch will have a predominantly lower frequency distribution than one of high pitch, but regardless of the actual speech characteristics there is always a significant part of the spectrum

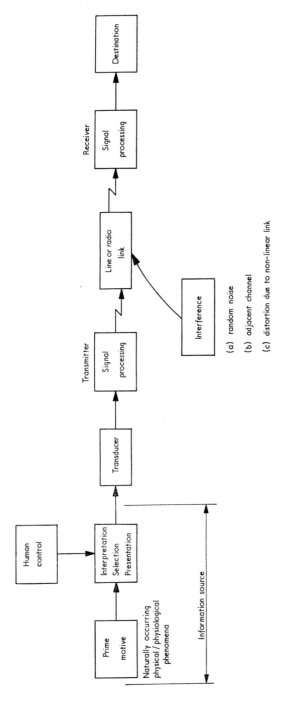

Fig. 1-2 Origination and telecommunication of a message.

within the range 300–3000 Hz. It will be seen in the chapters to follow that the range of frequencies occupied by the message must be limited to some specified bandwidth—a condition known as band-limiting and one which can be achieved by passing the electrical signal through a filter. For reasons of economy the bandwidth should be as small as possible, see Section 1–2, but a compromise must be reached between this requirement and the inevitable loss of information (or intelligibility).

In the case of speech the G.P.O. carried out a series of subjective tests to determine what this minimum bandwidth should be, and it was finally agreed and now internationally accepted* that 300–3400 Hz is sufficient for intelligibility, and that this also allows some of the characteristics, which are unique to a particular voice, to be communicated. Sometimes, for reasons of brevity, the bandwidth of a speech signal is referred to as 3 kHz rather than 300–3400 Hz.

A detailed account of speech characteristics can be found in Reference 1.

1–1.2 *Telegraphy*

The first form of telegraphy was invented by Samuel Morse in 1832 and consisted of the well-known 'dots' and 'dashes' code. Although this type of signalling is ideal for manual keying, it does not lend itself to mechanised systems, firstly because of the unequal element groups representing each character and secondly because two different elements, one three times the duration of the other, are required. Instead, teleprinters (or teletypewriters) use a code group of five equal-length elements to represent a keyboard character, and each element can have one of two possible states, referred to as either a mark (or 1) or a space (or 0)—i.e. it is a binary digit. Out of thirty-two (2^5) combinations of five binary digits, six are used for special purposes, such as carriage return, and the remaining twenty-six represent the alphabetic characters. In fact, this number is effectively doubled by using two of the 'special purpose' combinations to operate a 'letters' shift and a 'figures' shift in a manner similar to that of the ordinary typewriter.

The operation of a teleprinter communication link can be achieved in two ways. The first and simplest method is to transmit the electrical signals immediately they are formed, known as on-line working, but it suffers the disadvantage of having relatively long and irregular intervals between the characters. The second and more attractive arrangement is

* International Telegraph and Telephone Consultative Committee (CCITT) of the International Telecommunication Union (ITU).

1. *Communication System Engineering Handbook*, Editor D. H. Hamsher, Chapter 3 (McGraw-Hill, 1967).

to have the teleprinter working in an off-line mode. An electrical output is not required; instead the machine prepares the digital information in punched form on paper tape and the irregular time intervals that occur with manual operation of the keyboard are eliminated, since the characters are stored with uniform spacing. When the complete message, or a sufficient quantity of information, has been prepared in this way the electrical signals can be generated by passing the tape through a transducer known as a paper-tape reader. The off-line method also has the advantage of allowing the data to be regenerated at a faster rate than the output from the manually controlled teleprinter. Naturally there must be a drawback to such a system and, apart from the cost of additional equipment, there is an inevitable time delay between the origination of the message and its eventual transmission.

Let us consider the on-line system in more detail. The function of the receiving teleprinter is to sample each incoming digit and determine whether it is a mark or a space, to recognise a group of digits and then to print the appropriate character on a page of paper. The sampling rate is set by the speed of the teleprinter motor and ideally this should be exactly the same speed as the motor of the transmitter. In practice the motor speeds will always be slightly different and therefore the incoming data may be either sampled too quickly, some of the digits being sampled twice, or too slowly, which would result in some of the digits being undetected. To overcome this problem of synchronisation the receiver sampling is initiated at the commencement of each group of digits representing a character. This continual resetting of the sampling process is made possible by transmitting start and stop elements with each 5-digit group and is known as the 5-unit start-stop code. A description of the electro-mechanical operation is not within the terms of reference of this book, but details can be found elsewhere.[2,3]

The main criticism of on-line working is that the start and stop elements do not convey information of the message and are therefore to be regarded as redundant digits. An off-line mode of operation need not suffer this disadvantage. The 5-unit start-stop code for each character is stored, for example on paper tape, and the digits for transmission read out without the start and stop elements. At the receiving terminal, however, the digits are first stored and then read out with start and stop elements inserted. In such a system the transmitted digits are kept to a minimum, a desirable feature since the transmission time will also be kept to a minimum, but the read-out speed of the transmitting store and the read-in speed of the receiving store must be synchronised. Although

2. Chapter 2 of Reference 1.   3. *Telegraphy*, J. W. Freebody (Pitman, 1959).

[1–1]

this presents a problem it is by no means a prohibitive one. Modern techniques employ crystal-controlled oscillators (or clocks, as they are called) to generate the sampling pulses, and at the receiver the digit rate of the incoming signal is sensed and the frequency of the local clock adjusted for synchronism.

Next, let us look at the digit rate with a view to determining the frequency spectrum (or bandwidth) occupied by the signal. A typical speed of operation of European teleprinters produces information digits of 20 ms duration accompanied by a 20 ms start digit and a 30 ms stop digit (see Fig. 1–3). The duration of an information element defines the speed of operation and in this example they are generated at a rate of fifty per second and the telegraph speed is referred to as 50 bauds—the units being named after the French pioneer, Baudot. For data other than telegraphy the speed is quoted as the number of digits per second and, in the case of a binary signal, as the number of binary digits (or bits) per second.

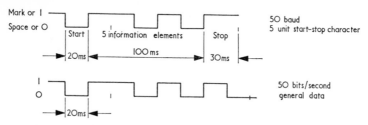

Fig. 1–3 Examples of binary signals with element duration of 20 ms.

Referring to the above example, we see that at 50-baud operation a complete character is generated in 150 ms, i.e. the character rate is 6·6 per second, and if the average word is made up of five characters plus a space, which also requires five digits plus start and stop elements, the average word speed is $(6·6 \times 60)/6 = 66$ words per minute. Examples of other operating speeds are listed in Table 1–1.

Consider a sequence of alternate marks and spaces, each of 20 ms duration (see Fig. 1–4). If the digits are perfectly rectangular with zero rise and fall times, the bandwidth of the signal would be infinite and consist of a line spectrum as defined in Example I–1, page 208. But since this is a digital signal it is only necessary for the receiver to determine whether a pulse represents a mark or space, the rise and fall characteristics being unimportant.† Theoretically it is possible to transmit this

---

† Although this statement will suffice for the present discussion, it is not strictly speaking accurate, since the distortion of a digit (or pulse) gives rise to intersymbol (or interdigit) interference, see Chapters 6 and 8.

Table 1-1

Typical Teleprinter Signal Characteristics

Nominal words per minute	Characters per second	Elements per character	Bauds (bits/s)	Start and information duration, ms	Stop duration, ms
66	6·67	7·5	50	20	30
75	7·5	10*	75	13·33	13·33
100	10·6	7·0	74·2	13·48	13·48
100	10·0	7·5	75	13·33	19·99
150	15·0	10*	150	6·67	6·67
600	60	10*	600	1·67	1·67
1200	120	10*	1200	0·83	0·83
2400	240	10*	2400	0·417	0·417

* 8 Information elements plus start and stop.

binary sequence by generating a 25 Hz sine wave—the positive and negative peak amplitudes representing the marks and spaces respectively—this, of course, is the fundamental component of the line spectrum. It will be appreciated that a typical message will give rise to a

Fig. 1-4 Periodic sequence of alternate marks and spaces.

random sequence of marks and spaces and not the simple alternating series considered above. The frequency spectrum occupied by such a non-periodic pulse train will not be made up of discrete components but will have a continuous distribution. At this stage, however, it is not desirable to become involved with the random signal and its continuous spectrum, and since we are only concerned with assessing the likely order of the bandwidth in terms of the digit rate, the above example of alternate marks and spaces serves as a useful illustration. The other extreme would be a sequence of all marks, or spaces, which would be represented as a steady d.c. signal. We may therefore conclude that the minimum baseband sufficient for transmitting the information of a random sequence of digits is numerically equal to half the digit rate. There is, of course, no theoretical bound on the maximum bandwidth that the signal can occupy but it will be shown, later in this chapter, that for economic reasons the bandwidth should be made as small as possible.

[1–1]

1–1.3 *Facsimile—Transmission of Still Pictures*

In facsimile systems pictorial information is scanned and converted into an electrical signal for transmission over a line or radio circuit and at the receiving terminal the information is reconverted into picture form. There are many techniques available for scanning and recording,[4] but in this book we are concerned only with establishing the bandwidth occupied by the electrical signals. Nevertheless, it is convenient to consider a simple form of scanning, see Fig. 1–5(*a*).

The picture to be transmitted is wrapped around a drum which is rotated by a lead screw and, as a result, a light source illuminating a small area scans along a helical path. The reflected light is sensed by a photo-electric cell which generates an electrical output fluctuating in sympathy with the picture pattern and varying in amplitude according to the light intensity.

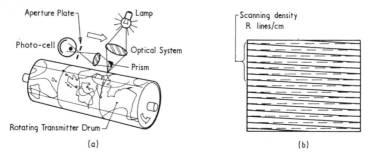

Fig. 1–5 (*a*) Scanning of a picture for facsimile transmission. (*b*) Unwrapped picture showing scanning lines.

The picture detail (or definition) conveyed in the signal will depend upon the pitch of the lead screw and the illuminated spot area. It is usual to arrange for adjacent areas, separated by one revolution of the drum, to touch each other, and in practice this results in a scanning density, R [or vertical definition, see Fig. 1–5(*b*)] of between 19 lines/cm and 50 lines/cm. Picture detail within an area illuminated by the spot may be lost, since the photo-electric cell will respond to the average intensity of the reflected light. As the spot scans along the helix, pattern changes separated by the diameter of the illuminated area will be detected by the cell. Take, for example, an alternating squared pattern as shown in Fig. 1–6(*a*). The corresponding electrical signal will be of the form shown in Fig. 1–6(*b*) and it would have the maximum peak-peak amplitude when the squares are alternating black and white. If the spot is scanning the picture at v cm/s and the pattern elements are similar in

4. Chapter 2 (pp. 67–84) of Reference 1.

dimensions to the spot size, then the frequency of the signal shown in Fig. 1–6(*b*) is $vd/2$ Hz, where d is the number of elements/cm.

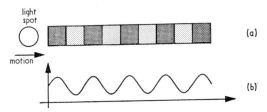

Fig. 1–6 (*a*) Alternating squared pattern scanned by light source. (*b*) Corresponding electrical signal.

It is to be noted that signals at this frequency will represent a horizontal definition that is the same as the vertical definition, i.e. $d = R$. There can, of course, be higher frequencies in the signal, caused by rapid pattern changes within an area covered by the spot, but the averaging of the reflected light by the photo-electric cell will cause these components to appear as a ripple with considerably reduced peak-peak amplitudes: for example, a rapid change from black to white will be conveyed as a signal corresponding to varying shades of grey.

In practice, and certainly for our purposes, the highest significant frequency in the electrical signal is that which gives a horizontal resolution similar to the vertical resolution, viz. $vd/2$ (or $vR/2$) Hz. In terms of the drum diameter D and speed N revolutions/s, the highest significant frequency is $\pi DNR/2$ Hz. Take, for example, a typical drum speed of 90 revs/minute, drum diameter 15·2 cm, and a scanning density of 38 lines/cm; then the highest significant frequency is

$$\frac{\pi \times 15\cdot 2 \times 90 \times 38}{60 \times 2} = 1360 \text{ Hz}$$

It is instructive to evaluate the total time required to scan a picture. Consider, for example, one with dimensions 45·7 cm × 56 cm. The time for 1 revolution of the drum is $1/N = 60/90$ seconds and the time to scan 1 cm in a direction along the axis of the drum is $R/N = (60 \times 38)/90 = 25\cdot 3$ seconds. Therefore the time to scan 56 cm is approximately 24 minutes.

The numerical values in the above example are typical of electro-mechanical scanning systems and we see that the bandwidth of the signal extends from d.c., corresponding to no pattern changes, to 1360 Hz. In other words, facsimile transmission of this kind has a bandwidth similar to speech and therefore it can be transmitted over a circuit that

[1–1]

is primarily designed for speech. Although not in common use, there are facsimile systems with electronic scanning methods permitting very much higher speeds of operation, but naturally the bandwidth of the signal would increase with increase in scanning speed.

1–1.4 *Television*

The essential difference between facsimile and television is that the former is concerned with the transmission of a still (or hard) picture, whereas television is concerned with moving pictures. The television camera scans the picture in a manner similar to the light source scanning the rotating drum in the facsimile system, see Fig. 1–5 (*b*), but for television the dashed lines are a real part of the scanning operation and referred to as the flyback.

The number of complete pictures that must be transmitted per second is determined by the human characteristics of persistence of vision. The eye cannot detect a flicker or abrupt discontinuity from picture to picture if they are changing at a rate of 20 or more per second. In the U.K. the moving picture is made up of 25 complete pictures per second with either 405 or 625 lines defining the vertical resolution. As a result of subjective tests during the early days of television, the transmission consists of 50 half-pictures (or frames) per second each of which is made up of $202\frac{1}{2}$ or $312\frac{1}{2}$ lines, corresponding to 405- or 625-line systems respectively, with alternate frames interlaced. If we consider similar horizontal and vertical definitions, as for facsimile, then the highest significant frequency of the electrical signal is

$$\frac{405 \times 405 \times 25}{2} \times \frac{4}{3} \approx 2\cdot7 \text{ MHz}$$

The factor 4/3 is the aspect ratio to account for the picture not being square.

1–2 Line Communication—The Linear Channel

The simplest method of conveying an electrical signal from a transmitting terminal to a distant receiving terminal is by a pair of wires. The channel (or circuit) can be regarded as linear when the following conditions apply:

(i) the output voltage and current are directly proportional to the input voltage and current, and

(ii) the output corresponding to two input signals applied simultaneously is the sum of the two outputs obtained had the inputs been applied individually—the Superposition Theorem.

The circuit exhibits a non-linear characteristic when sinusoidal components, other than those applied at the input, appear at the output, but for the purpose of our present discussion we shall consider the channel to be linear.

On page 1, it was shown that signals can be classified as being either analogue or digital. For the former it is essential that the received signal closely resembles the transmitted signal, and that its amplitude (or power) should be very much larger than the amplitude (or power) of the accompanying noise or interference. Digital signals, on the other hand, can tolerate considerably more interference, see Chapter 8. However, for both types it is essential to know the transmission characteristics of the line.

Consider first of all a four-terminal (or two-port) network connecting a generator to a load, see Fig. 1–7(a). The transmission loss can be

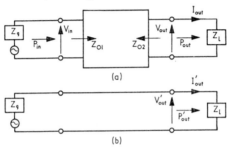

Fig. 1–7 (a) Four-terminal network connected to a generator and load. (b) Network replaced by straight-through connections.

expressed as the ratio of the input and output powers, P_{in}/P_{out}, or as the logarithm of this ratio, i.e. $\log_{10}(P_{in}/P_{out})$.* Although this quantity is dimensionless, it is referred to in 'units' of bels—named after Graham

* It is convenient to refer to the logarithm of the power ratio since the network under consideration is often one of several in tandem. If the loss characteristic, or gain characteristic if active devices are included, of each of the networks is known, the overall transmission gain or loss is obtained by simple addition instead of multiplication had just the power ratios been used. For example, in the following system:

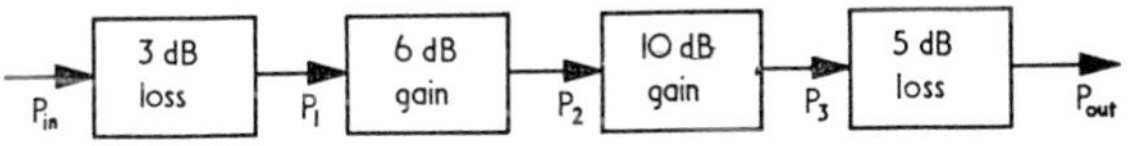

Fig. 1–8.

the overall transmission loss or gain

$$\frac{P_{out}}{P_{in}} = \frac{P_1}{P_{in}} \times \frac{P_2}{P_1} \times \frac{P_3}{P_2} \times \frac{P_{out}}{P_3}$$

or the overall gain in dB, by addition, $= -3 + 6 + 10 - 5 = 8\text{dB}$, i.e. the power ratio $P_{out}/P_{in} = \text{antilog}(0.8)$.

[1–2]

Bell—to denote that the base of the logarithm is 10. For ease of handling, it is usual to use a unit which is one-tenth the size of the bel, namely the *decibel* (dB). Thus for the four-terminal network the

$$\text{transmission loss} = 10 \log_{10}(P_{in}/P_{out}) \text{ dB} \qquad (1\text{-}1)$$

Usually, Z_g and Z_l are resistive, and the network is designed for $Z_{01} = Z_g$ and $Z_{02} = Z_l$ over the frequency band occupied by the signal. Furthermore, when the network is symmetrical, the four impedances are equal and resistive. Under these conditions the

transmission loss

$$= 10 \log_{10}(|V_{in}|/|V_{out}|)^2 = 20 \log_{10}(|V_{in}|/|V_{out}|) \quad (1\text{-}2)$$
$$= 10 \log_{10}(|I_{in}|/|I_{out}|)^2 = 20 \log_{10}(|I_{in}|/|I_{out}|)$$

The moduli of the voltages and currents are used since Eq. (1-2) is derived from the transmission loss defined in terms of the input and output powers. So far, no reference has been made to the phase relationship between V_{out} and V_{in}; we shall see that this is an important factor and must be known. It should be mentioned for the benefit of the student who is not familiar with basic network theory that although Z_{01} and Z_{02} are resistive the input and output voltages and currents need not be in phase. This is certainly true for a transmission line which is a four-terminal network in distributed form; the propagation delay, which is a function of the length of line, gives rise to a phase difference between the input and output signals.

The transmission characteristic (or *transfer function* as it is more commonly known) defining both attenuation and phase relationships is given by

$$\left.\begin{array}{c} \dfrac{V_{in}}{V_{out}} = \dfrac{|V_{in}|}{|V_{out}|} \exp(j\phi) \\[2em] \text{or as} \qquad\qquad \\[1em] \dfrac{V_{out}}{V_{in}} = \dfrac{|V_{out}|}{|V_{in}|} \exp(-j\phi) \end{array}\right\} \qquad (1\text{-}3)$$

where ϕ is the phase difference between V_{in} and V_{out}—it assumes that the input is a single component, i.e. a sine wave. Except for the case of a network being made up of 'pure' resistors, i.e. an attenuator, the transfer function varies with frequency, and to denote this dependence the output/input ratio is represented as

$$\frac{V_{out}}{V_{in}} = H(j\omega) = A(\omega)\exp[-j\phi(\omega)] \qquad (1\text{-}4)$$

where $A(\omega)$ represents the attenuation (or amplitude)-frequency response and $\phi(\omega)$ the phase-frequency response. In order to simplify the appearance of mathematical expressions, the transfer function may be referred to as $H(j\omega)$ denoting that in general it is complex and a function of frequency. Example I–6, page 224, illustrates its evaluation for three simple networks.

The transmission loss given by Eq. (1–1) has been defined in terms of the input and output power ratio. An equally valid and, in fact, well-established method of analysing networks begins by defining the input and output current ratio as

$$\frac{I_{in}}{I_{out}} = \exp \gamma = \exp(\alpha + j\beta) \qquad (1\text{–}5)$$

where γ is known as the propagation constant, α defines the attenuation and β the phase relationship. It follows that

$$\frac{|I_{in}|}{|I_{out}|} = \exp \alpha$$

or expressed as a logarithm to the base e, the attenuation of the current is given by

$$\alpha = \log_e \left(\frac{|I_{in}|}{|I_{out}|} \right) \text{ nepers} \qquad (1\text{–}6)$$

and although this quantity is dimensionless it is defined in 'units' of *nepers* to indicate that natural logarithms are used. When the condition $Z_g = Z_l = Z_{01} = Z_{02}$ applies,

$$\left. \begin{aligned} \alpha &= \log_e \left(\frac{|V_{in}|}{|V_{out}|} \right) \\ &= \tfrac{1}{2} \log_e \left(\frac{P_{in}}{P_{out}} \right) \end{aligned} \right\} \qquad (1\text{–}7)$$

A comparison of Eqs. (1–2) and (1–6) shows that the

attenuation in dB $= 8.686 \times$ *attenuation in nepers.*

Eqs. (1–3) and (1–5) indicate that ϕ and β are identical and can be expressed in radians or degrees. Throughout this book the transfer function of Eq. (1–4) is used and voltage, current and power ratios expressed in decibels.

So far, the attenuation characteristics of a network have been defined as a transmission loss from a definition of either the power ratio or the current ratio. A second method considers the reduction in power in the load due to the insertion of the network. Referring to Figs. 1–7 (*a*) and 1–7 (*b*) we have by definition the

$$insertion\ loss = 10 \log_{10} \left(\frac{P'_{out}}{P_{out}} \right) \text{ dB} \qquad (1\text{–}8a)$$

$$= 20 \log_{10} \left(\frac{|I'_{out}|}{|I_{out}|} \right) \text{dB} \qquad (1\text{–}8b)$$

Note that Eqs. (1–8a) and (1–8b) can be established regardless of

[1–2]

whether Z_g, Z_l, Z_{01}, Z_{02} are equal or not, since the two currents or powers are associated with the same impedance, namely Z_l. When the source and load impedances are equal and matched to the network, the insertion loss and transmission loss are identical.

When a linear circuit consists of resistors, inductors and capacitors it is said to be made up of lumped components as distinct from the distributed type of network, such as a pair of wires forming a transmission circuit. In the latter case the transmission loss, insertion loss and transfer function will depend upon the length of the wire and it is therefore convenient to quote these characteristics on a per unit length basis.

Suppose we wish to transmit electrical signals over a pair of twisted wires. Two questions will come to mind: firstly, can all types of signals be transmitted and, secondly, what will be the maximum length of the circuit? Unfortunately a simple straightforward answer is not possible, but the following explanation allows the additional influencing factors to be introduced.

The maximum length of the line will depend upon the transmitted signal power, line attenuation and the minimum acceptable signal-to-noise power ratio* at the receiver. The transmitted signal power must be of the order of watts rather than kilowatts to avoid overheating the wire and having a high level of crosstalk between close-spaced pairs of wires carrying independent messages. The line attenuation must be kept as small as cost will allow—the larger the diameter of the wire the lower the attenuation but the higher the cost. However, before this statement can be discussed we must look at the transmission loss characteristics of some typical lines.

The inductive and capacitive properties of the line cause the transfer function to be frequency dependent. The transmission loss and phase-frequency characteristics of a pair of 20 SWG polyethylene insulated wires are shown in Fig. 1–9. If a 3 kHz speech signal is transmitted over the channel the higher frequency components would suffer more attenuation than the lower frequency components, and for television signals the amplitude distortion would be considerably worse.

In the early days of line communication it was realised that the transmission loss-frequency characteristic can be made flat by adding inductance to the line. Theoretically if this could be achieved in a distributed form, e.g. wrapping the wire with mumetal tape, there is no limit on the bandwidth over which this compensation is possible. In practice a distributed addition of inductance is uneconomical and instead coils

* The minimum permissible signal-to-noise power ratio depends upon the particular requirement; for example, commercial telephony can tolerate 25–30 dB, whereas for high-quality sound reproduction it is desirable to have a ratio in excess of 50 dB.

must be inserted at intervals along the line—this is known as lumped loading and the modification to the transmission loss is illustrated in Fig. 1–9. At very low frequencies the loading is effectively in distributed form, but at frequencies around 3 kHz the coils together with the capacitance between the lines form a low-pass filter which results in a rapid increase in attenuation at high frequencies. This method of minimising the amplitude distortion is particularly effective for speech transmission and is used on circuits of up to thirty miles in length with the loading coils spaced at intervals ranging from 0·9 to 1·83 km. Compensation over a band sufficiently wide for television transmission cannot be achieved and a low-loss cable is essential.

The second frequency-dependent factor is the time delay between the signal leaving the transmitter and arriving at the receiver. If this were constant over the range of frequencies occupied by the signal, there would be no problem, but, unfortunately, some components arrive sooner than others, or in other words the instantaneous phase relationships at the receiver are different from those of the original signal at the transmitter. Phase distortion, as it is known, is not troublesome on speech circuits because the ear is insensitive to this phenomenon. It is somewhat ironical that the loaded line suitable for speech transmission has a remarkably constant propagation velocity—and hence delay time—over the 3 kHz bandwidth! For television purposes phase distortion must be kept to a minimum and this is accomplished by terminating the line with an amplifier, known as an equaliser, which compensates for both amplitude and phase distortion. When long-distance transmission is required the amplifiers (or repeaters) are included at intervals along the line, the spacing being determined by the line characteristics and the overall bandwidth of the signal; for example, repeaters at intervals of 9·86 km (6·125 miles) are used for transmitting 405-line television signals over a 9·5 mm (0·375 inch) diameter coaxial cable.

If the time delay, T_d, is to be independent of frequency, the phase characteristic of the transfer function must be

$$\phi(\omega) = -\omega T_d \text{ radians}$$

where ω is the angular frequency in radians/second and the negative sign indicates that the output is lagging the input. A linear phase-frequency characteristic is, unfortunately, physically unrealisable except for the transmission line with distributed loading. The high cost of such a line makes it unattractive and it is cheaper to use a channel with non-linear phase characteristic together with a phase compensating circuit, i.e. an equaliser.

[1–2]

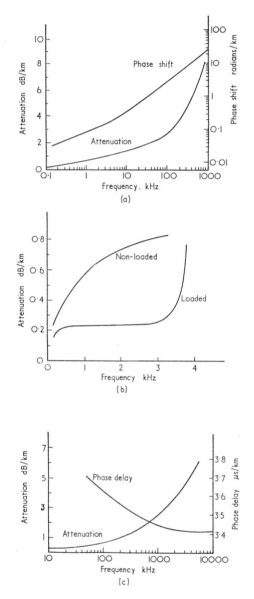

Fig. 1–9 Typical attenuation-frequency and phase-frequency characteristics of (a) 20 SWG (19 AWG), 5·6 kg/km (20 lb/mile) polyethylene insulated cable pairs; (b) the above line loaded with 88 mH inductors spaced at 1·83 km intervals, and (c) 9·5 mm (0·375 inch) diameter coaxial cable.

Let us consider transmitting a speech signal over a line circuit connecting two points hundreds of kilometres apart. The loaded line would be suitable provided that the overall attenuation could be tolerated. Over a long circuit this would be highly unlikely and an acceptable signal-to-noise power ratio at the receiving terminal could only be achieved by inserting amplifiers at intervals along the line, say, every eighty kilometres. A second speech channel would require the same amount of line plant, i.e. the wire, loading coils and repeaters. This leads us to consider the simple case of connecting two exchanges in such a way as to permit simultaneous transmission of many messages. If the distance of separation is less than forty kilometres, or thereabouts, a large number of individual circuits provides the cheapest method of interconnection, but for longer distances a considerable saving in copper can be realised by transmitting a number of speech signals over a single circuit. At the transmitting terminal the signals, each occupying a 3 kHz baseband, are transposed in frequency in such a way that their characteristics, apart from the absolute frequencies, are not altered. In practice 12 channels, known as a *Basic Group*, form a composite signal occupying an overall bandwidth of 48 kHz with each channel having a 4 kHz 'slot' (see Fig. 1–10). The process of transposing a baseband signal to some other part of the frequency spectrum can be classified under the general title of *modulation*, and the sharing of a given bandwidth, 48 kHz in the example, between a number of independent channels is known as *frequency division multiplex* (FDM). A line suitable for transmitting the multichannel signal must be capable of handling all frequencies up to 48 kHz and therefore the loaded line would not be acceptable. However, the same line unloaded could be used in conjunction with amplifiers serving as equalisers and repeaters. It is common practice to arrange

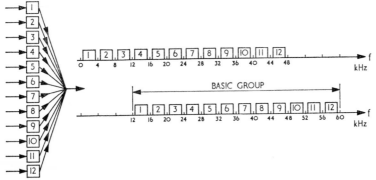

Fig. 1–10 Formation of a FDM signal of 12 telephone channels.

[1–2]

the channels to occupy a frequency band from 12 to 60 kHz in place of 300 Hz to 48 kHz (see Fig. 1–10); the ratio of the highest to lowest frequency is therefore considerably smaller and this is advantageous for the design of the amplifiers. A number of Basic Groups can be placed side by side in the frequency domain with the second group in the band 60–108 kHz, the third from 108–156 kHz and so on; five of these make up a *Supergroup*. Still further multiplexing is possible with five Supergroups arranged in the frequency band 60–1300 kHz (see Fig. 1–11)—note that the 12–60 kHz Basic Group has been removed to make the ratio of the maximum to minimum frequencies smaller.

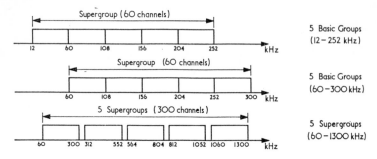

Fig. 1–11 Multiplexing of telephone channels.

It will be realised that a multichannel signal consisting of a number of Basic Groups or Supergroups cannot be transmitted very far over a pair of 20 SWG wires, since the attenuation of the high frequencies would be prohibitively high and equalisation virtually impossible to achieve. A low-loss coaxial cable can provide the necessary bandwidth, and one in common use has a 9·5 mm ($\frac{3}{8}$ inch) inner diameter of the outer conductor with a 2·54 mm diameter inner conductor. A capacity of 960 speech channels, i.e. 16 Supergroups in a 4 MHz bandwidth, can be realised when repeaters spaced every 9·68 km (6·125 miles) are used. The same cable can accommodate 2700 channels, contained in a 12 MHz bandwidth, when the repeater spacing is 4·5 km (2·8 miles)—this, in fact, is the largest capacity line system in present day use.

At the receiving terminal the FDM signal undergoes a reverse translation process, known as *demodulation*, to separate the speech channels and restore each one to the original common baseband of 300–3400 Hz.

Multichannelling of television signals is not feasible because of the very large bandwidth required for each channel and the practical difficulties associated with their transposition. There are, however, 12 MHz systems which allow a television channel and many telephone channels to form a FDM signal. It is interesting to note that a wideband 4 MHz coaxial line system can handle 960 3-kHz speech channels or one television signal.

The main purpose of the preceding discussion has been to introduce the concept of frequency transposition (or modulation) and to show, with examples of analogue signals, how it may be applied to line communication systems. It is not appropriate to discuss digital data transmission at this stage, the relevant points being developed in later chapters.

We can now return to the question posed on page 14 regarding the maximum length of the line and the types of signals than can be handled. For speech we have seen that the transmission of one signal over a long distance would be an hypothetical consideration, since in practice there would be a need to convey the information of many independent messages. Economic reasons dictate the use of separate channels for short distances and multichannelling on a low-loss coaxial line for long distances, with the breakpoint occurring at about forty kilometres. A circuit carrying a FDM signal can be any desired length provided that repeaters are used at regular intervals. This also applies to the transmission of television signals except that equalisation, to prevent excessive phase distortion as well as amplitude distortion, is essential.

1–3 Radio Communication

It is well known that when a quantity of electro-magnetic energy is set up in space it cannot remain at rest, but must travel as a wave with a velocity, c, of approximately 3×10^8 m/s until all the energy is dissipated. Suppose that an observer at a fixed point detects the wave motion and sees that the field pattern varies sinusoidally with time, at a frequency f cycles/s (or Hertz). The time for one cycle is $1/f$ seconds and, since the velocity of propagation is known, the spatial distribution can be defined in terms of a wavelength λ, i.e. the distance between points of equal field intensity is given by $\lambda = c/f$ (or $c = f\lambda$).

The frequency range of the electro-magnetic spectrum and the sub-classifications are illustrated in Fig. 1–12. The division into sections labelled cosmic rays, X-rays, light waves and radio waves is made because the propagation characteristics differ considerably from one part of the spectrum to another. Visible light, for example, is not easily diffracted, i.e. it casts shadows, and it cannot pass through opaque objects. Radio waves, however, can be diffracted and can pass through non-ferromagnetic opaque bodies, although they will be, to some extent, attenuated. It is these factors plus the ease of frequency translation (or modulation) that have made the lower frequencies of the e-m spectrum,

[1–3]

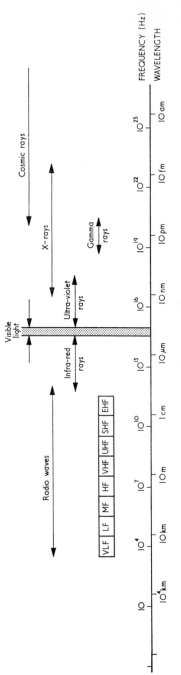

Fig. 1-12 The electromagnetic spectrum.

Table 1-2

Classification of the Radio Wave Spectrum

Band number	Band designation*	Frequency range	Wavelength range	Metric classification	Main services	Transmission range (salient propagation characteristics)
4	VLF (very low frequency)	3–30 kHz	100–10 km	Myriametric waves	Navigation	World-wide
5	LF (low)	30–300 kHz	10–1 km	Kilometric waves	Telegraphy	
6	MF (medium)	300–3000 kHz	1 km–100 m	Hectometric waves	Regional broadcasting; shipping	Limited range of hundreds of km
7	HF (high)	3–30 MHz	100–10 m	Decametric waves	Civil and military point-to-point; amateur broadcast;	World-wide (by ionospheric and ground reflection)
8	VHF (very high)	30–300 MHz	10–1 m	Metric waves	as for HF	Optical line of sight (negligible diffraction)
9	UHF (ultra high)	300–3000 MHz	1 m–10 cm	Decimetric waves	as for HF; long range radar	as for VHF, plus distances up to 1000 km by tropospheric scatter
10	SHF (super high)	3–30 GHz	10–1 cm	Centimetric waves	Radar; satellite systems	Optical line of sight
11	EHF (extra high)	30–300 GHz	1 cm–1 mm	Millimetric waves	Research	Optical line of sight (high atmospheric attenuation)
12	—	300–3000 GHz	1–0·1 mm	Decimillimetric waves	—	

k(kilo) = 10^3, M(Mega) = 10^6, G(Giga) = 10^9.

* It is general practice to refer to the band designation rather than the metric classification except for the millimetric range.

10^4 to 10^{10} Hz, particularly suitable for communicating over long distances.

Apart from the propagation conditions varying considerably over the e-m spectrum there are marked differences within the radio wave section. The treatment of this subject requires a textbook devoted solely to radio-wave propagation, and hence it cannot be discussed adequately in a few pages of this chapter. Rather than not mention the subject at all, it is intended that the subdivision of the radio frequencies together with a summary of the main services and transmission ranges, detailed in Table 1–2, will serve as a useful guide.

The allocation of channels which might cause interference to other services, particularly those belonging to another administration, is the responsibility of the International Frequency Registration Board (I.F.R.B.) of the International Telecommunication Union (I.T.U.). Frequency assignment within the HF band and for satellite channels in the SHF region are particularly important. In the VHF and UHF spectra a channel may be allocated to a number of different administrations throughout the world, since the transmission range is restricted to what is virtually line of sight. The classification of the broadcast band has been reached by international agreement, but the allocation of channels is left to the governing body responsible for the telecommunications of that particular country—the U.K. channel designation and allocation is detailed in Table 1–3.

Table 1–3
Broadcast Band Classification

Band	Frequency range	U.K. channel designation	U.K. service
I	41 – 68 MHz	1 to 5	BBC 1 (405 tv)
II	87·5 – 100 MHz	—	BBC (VHF/FM sound)
III	174 – 216 MHz	6 to 13	ITA (405 tv)
IV*	470 – 582 MHz	21 to 34 ⎱ 44 channels, each	BBC 2 (625, colour)
V*	614 – 854 MHz†	39 to 68 ⎰ 8 MHz bandwidth	ITA 2nd channel (625, colour)

* International allocation of the UHF band is 470–960 MHz, but European services have only 470–582 MHz and 614–960 MHz.

† U.K. allocation for Band V.

Finally, an explanation of the commonly used term 'microwaves' is necessary. The British Standards Institution recommend that electromagnetic waves of frequencies higher than 1 GHz should be classified as such, but another definition uses the term to indicate that transmission line and waveguide techniques are employed in the equipments. The latter interpretation means that the microwave region is extended downwards to include nearly all of the UHF band. Table 1–4 lists the classification of microwaves according to a radar band 'letters' designation. Extreme caution should be exercised when using these notations, since they have not been internationally standardised.

Table 1–4
Classification of Microwaves

Frequency range, GHz	U.K. band designation	Frequency range, GHz	American band designation
0·375– 1·50	L	1 – 1·88	L
1·5 – 3·75	S	2·35– 4·175	S
3·75 – 6·00	C	3·6 – 8·65	C
6 – 11·5	X	7·5 – 13·2	X
11·5 – 18	J	11 – 19·2	KU, P
18 – 30	K	16 – 28	K
30 – 47	Q	33 – 50	Q

2
ANALOGUE MODULATION

2–1 Amplitude Modulation

The need to transpose a baseband signal to some other part of the spectrum is essential for radio transmission and highly desirable if the economic advantages of wideband coaxial line systems are to be realised. The process of transposition is called *modulation* and in its simplest form the amplitude variations of the baseband signal are made to occur within the same bandwidth, but higher in the spectrum, e.g. a 300–3400 Hz speech signal may be translated in frequency to, say, 12300–15400 Hz without any alteration to the amplitude variations.

Let us see how this may be achieved. Firstly, it is reasonable to assume that the baseband signal must be combined with a sinusoidal component whose frequency is in the region where the translation is to be effected. The combining operation must be performed by a non-linear circuit* since the input and output frequencies are to be different. Theoretically, the simplest, and mathematically most convenient, non-linear operation is a straightforward multiplication of the baseband signal $m(t)$ and the higher frequency sinusoid, $E_c\cos\omega_c t$. The output $e(t)$ from the multiplier can be expressed as either

$$e(t) = m(t)\cos\omega_c t \qquad (2\text{--}1)$$

or
$$e(t) = km(t)E_c\cos\omega_c t \qquad (2\text{--}2)$$

where the constant k has the dimensions of (volts)$^{-1}$ and must be included to make the equation dimensionally correct. Eq. (2–1) represents an output that is independent of the peak amplitude of the sinusoid; the validity of this statement is established on page 49 in the treatment of a practical form of multiplier. For the present analysis either of the two equations could be used, but reference is made only to Eq. (2–1). A typical modulating signal and modulated output corresponding to this equation are shown in Fig. 2–1.

We now wish to establish the frequency spectrum of this signal to see if the baseband information has been transposed in the predetermined way. The modulating signal $m(t)$ is invariably of a complex nature. Referring again to speech as an example, the information is

* See page 10.

contained in the range 300–3400 Hz and, in fact, the spectrum is continuous, i.e. discrete sinusoids do not exist, which implies that the signal variations in the time domain are non-periodic.* A rigorous determination of the frequency spectrum corresponding to Eq. (2–1) requires an

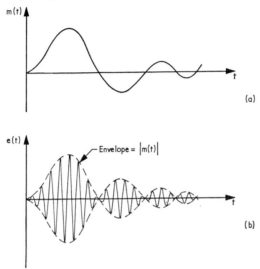

Fig. 2–1 (*a*) Typical modulating signal. (*b*) Output due to multiplication with a sinusoid cos $\omega_c t$.

understanding of Fourier transform techniques which are dealt with in Appendix I. Whilst it is essential for the reader to be familiar with these methods if this work and that of the following chapters are to be fully assimilated, a simpler and more instructive approach may be adopted by considering $m(t)$ to be made up of only a few sinusoidal components: for example, assume that

$$m(t) = E_1\cos\omega_1 t + E_2\cos\omega_2 t + E_3\cos\omega_3 t$$

The output is given by

$$e(t) = (E_1\cos\omega_1 t + E_2\cos\omega_2 t + E_3\cos\omega_3 t)\cos\omega_c t$$

$$= \frac{E_1}{2}[\cos(\omega_c+\omega_1)t + \cos(\omega_c-\omega_1)t]$$

$$+ \frac{E_2}{2}[\cos(\omega_c+\omega_2)t + \cos(\omega_c-\omega_2)t]$$

$$+ \frac{E_3}{2}[\cos(\omega_c+\omega_3)t + \cos(\omega_c-\omega_3)t]$$

* For a signal to contain information it must be non-periodic, otherwise the receiver would be able to anticipate with complete certainty all future events.

[2–1]

and a plot of the frequency spectrum is shown in Fig. 2–2(b). It can be seen that $e(t)$ consists of two groups, each of three components, symmetrically positioned about the frequency f_c. Since the non-linear operation has not caused any interaction between the components of $m(t)$ we may assume that a complex baseband signal within the range ω_1 to ω_3 will produce an output signal as shown in Fig. 2–2(d). The components from $(\omega_c+\omega_1)$ to $(\omega_c+\omega_3)$ constitute the *upper sideband* (USB) and the mirror image from $(\omega_c-\omega_3)$ to $(\omega_c-\omega_1)$ make up the *lower sideband* (LSB). The baseband, ω_1 to ω_3, will be referred to as B_m and in general ω_1 may extend to zero frequency, see Fig. 2–2(e). Having

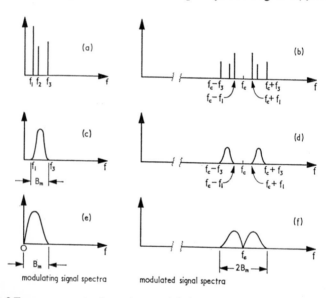

Fig. 2–2 Frequency spectra for various modulating and modulated signals. (a) and (b) three sine waves; (c) and (d) an aperiodic signal; (e) and (f) an aperiodic signal with
$$f_1 \rightarrow 0.$$

established the requirement of a non-linear circuit, it may seem paradoxical to refer to amplitude modulation as a *linear* process. But this is common practice and its justification relies on the fact that a component in the baseband gives rise to one component only in a sideband. *Non-linear* modulation processes, on the other hand, do not exhibit this one-to-one correspondence (see page 36).

Returning to the initial requirement of transposing the baseband signal to another part of the spectrum, we see that this is achieved by passing the signal, defined by Eq. (2–1), through a filter to remove the lower sideband. The technique is illustrated in Fig. 2–3. The amplitude of the *single sideband* (SSB) output differs from that of the baseband signal by a factor 2, but the relative amplitudes of the signal components before and after the multiplication process are the same.

In the design of communication systems, filtering of signals always presents a problem and we may therefore ask, 'Is it necessary to transmit the information in single sideband form or can the double sideband (DSB) signal be used?' The short answer is that both forms are suitable, since it is possible to recover the baseband information by demodulating either the SSB or DSB signals; a qualification of this statement is to be found in section 3–2, page 54, under Demodulation.

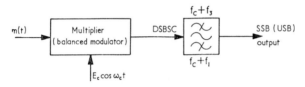

Fig. 2–3 Generation of a SSB signal.

A third form of modulation consists of upper and lower sidebands plus a component at the centre frequency f_c. Consider the block diagram of Fig. 2–4. The output of the system is

$$e(t) = E_c\cos\omega_c t + m(t)\cos\omega_c t$$

$$= [E_c + m(t)]\cos\omega_c t \qquad (2\text{–}3)$$

and is illustrated in Fig. 2–5. We see that the inserted component acts as a carrier and that the envelope of the waveform is a replica of $m(t)$.

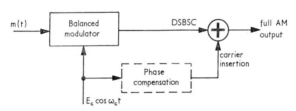

Fig. 2–4 Generation of a full AM signal.

For this reason, the signal is said to be the result of *full amplitude modulation* (full AM) and consists of a carrier whose peak-to-peak amplitude varies in direct proportion to the *modulating signal m(t)*, a condition that can only be realised if $|m(t)| \leqslant E_c$. In practice, it is possible to exceed this limit, but the demodulated signal would be distorted. A comparison

[2–1]

of Eq. (2–1) and Fig. 2–1 with Eq. (2–3) and Fig. 2–5 shows that the output from the multiplier (or balanced modulator)* is a double sideband suppressed carrier (DSBSC) signal and that there is no limit to the *depth of modulation.*

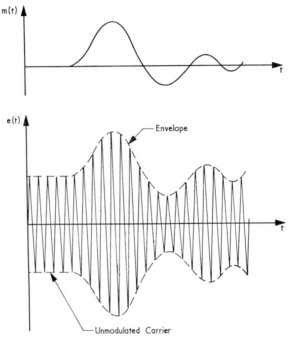

Fig. 2–5 Baseband and full AM waveforms.

It must be made clear that Eq. (2–3) assumes phase coherence between the modulator output and the added carrier. Strictly speaking the former should be written as $m(t)\cos(\omega_c t + \theta)$ where θ accounts for the unavoidable phase difference between input and output of any practical form of modulator. Phase compensation would have to be included in the carrier insertion circuit to satisfy the requirement of phase coherence. If the carrier and DSBSC signal are not in phase, the resultant modulation would be different from the full AM definition of Eq. (2–3)—a quadrature condition is dealt with on page 41.

The reason for introducing a carrier to form a full AM signal will not be apparent until the process of demodulation has been studied; however, it will be realised that its inclusion is not essential for the transmission of the message information, since the sidebands are ade-

* This term implies that a component at the carrier frequency does not appear in the output.

quate. If a transmitter has a total available power P_T, then some of this must be apportioned to the carrier and may be regarded as a loss of power so far as the generation of the sidebands is concerned. It is therefore of interest to determine the proportion of carrier power to sideband power. From Eq. (2–3) the normalised signal power, i.e. the power developed in a 1 ohm resistor, is

$$P_T = \overline{e^2(t)} = \frac{E_c^2}{2} + \frac{\overline{m^2(t)}}{2}$$

where $E_c^2/2$ is the mean carrier power P_c. From the above comments, whilst it may be concluded that $\overline{m^2(t)}$ should be made as large as possible, since the sidebands contain the message information, it is also necessary to ensure that $|m(t)_{max}| \not> E_c$; otherwise overmodulation would occur.

A continuation of a general analysis would require a knowledge of the statistical properties of the modulating signal in order that a relationship between $m(t)$ and $\overline{m^2(t)}$ may be established—the reader will appreciate the full implication of this statement later in the book. Since this is an introduction to amplitude modulation, we shall first consider $m(t)$ to be a single sinusoidal component, $E_m \cos\omega_m t$, rather than a complex signal. Thus Eq. (2–3) becomes

$$e(t) = (E_c + E_m\cos\omega_m t)\cos\omega_c t$$
$$= E_c(1 + m_a\cos\omega_m t)\cos\omega_c t \qquad (2\text{--}4)$$
$$= \underset{\text{carrier}}{E_c\cos\omega_c t} + \underset{\text{LSB}}{\frac{m_a E_c}{2}\cos(\omega_c - \omega_m)t} + \underset{\text{USB}}{\frac{m_a E_c}{2}\cos(\omega_c + \omega_m)t}$$

where $m_a = E_m/E_c$ and is known as the *modulation index*. To avoid overmodulation $m_a \leqslant 1$ (when $m_a = 1$, 100% modulation is said to exist). The signal power is

$$P_T = \overline{e^2(t)} = \underset{\text{carrier}}{\frac{E_c^2}{2}} + \underset{\text{LSB}}{\frac{m_a^2 E_c^2}{8}} + \underset{\text{USB}}{\frac{m_a^2 E_c^2}{8}}$$
$$= P_c + P_{LSB} + P_{USB} \text{ say.}$$

If m_a is set to its maximum value of unity, then

$$P_{USB} + P_{LSB} = \tfrac{1}{2}P_c$$

which means that only one-third of the total power is being used to convey the message information. In the case of a DSBSC signal, all the power is conveyed in the sidebands, and it therefore remains to be seen whether or not the power utilisation of the full AM form can be justified (see section 3–2).

[2–1]

Next consider the modulating input to be a band-limited periodic signal

$$m(t) = \sum_{n=-m}^{m} c_n \exp(jn\omega t) \qquad (2\text{-}5)$$

where $\omega = 2\pi/T$ and T is the time for a complete period of the signal (see Appendix I). The coefficient c_n can be complex, but since $m(t)$ is always a real function, then c_n must equal c^*_{-n} where the asterisk denotes the complex conjugate (see page 213). The exponential term $\exp(jn\omega t)$ defines the instantaneous phase of the nth component whose peak amplitude is $|c_n|$. The two-sided frequency spectrum which comes about because of the complex form of representation is shown in Fig. 2–6(*a*). The equally valid and in fact physically realizable, one-sided representation can be derived by substituting $c_n = |c_n| \exp(j\theta_n)$ in the above equation, i.e.

$$m(t) = \sum_{n=-m}^{m} |c_n| \exp[j(n\omega t) + \theta_n)]$$

$$= 2\sum_{n=1}^{m} |c_n| \cos(n\omega t + \theta_n) \qquad (2\text{-}6)$$

(See Fig. 2–6(*c*); cf. Eq. (i–15), page 218.)

The DSBSC form of modulated signal is given by

$$e(t)\bigg|_{\text{DSBSC}} = \sum_{n=-m}^{m} c_n \exp(jn\omega t)\tfrac{1}{2}[\exp(j\omega_c t) + \exp(-j\omega_c t)]$$

$$= \tfrac{1}{2}\sum_{n=-m}^{m} c_n \exp[j(\omega_c t + n\omega t)] + \tfrac{1}{2}\sum_{n=-m}^{m} c_n \exp[j(-\omega_c t + n\omega t)] \qquad (2\text{-}7)$$

and is illustrated in Fig. 2–6(*b*). The one-sided frequency spectrum is given by

$$e(t)\bigg|_{\text{DSBSC}} = 2\sum_{n=1}^{m} |c_n| \cos(n\omega t + \theta_n)\cos\omega_c t$$

$$= \sum_{n=1}^{m} |c_n| \cos[(\omega_c + n\omega)t + \theta_n]$$

$$+ \sum_{n=1}^{m} |c_n| \cos[(\omega_c - n\omega)t + \theta_n] \qquad (2\text{-}8)$$

and is shown in Fig. 2–6 (*d*).

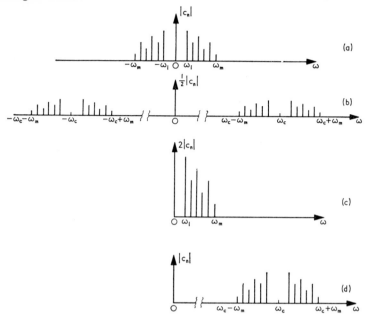

Fig. 2–6 Periodic modulating signal and DSBSC spectra. (*a*) and (*b*), two-sided frequency representation; (*c*) and (*d*), one-sided frequency representation.

The treatment of a non-periodic modulating signal is dealt with in Appendix I on page 235. The reader who is meeting this subject for the first time would be well advised to defer this study until more of the book has been covered.

Nowadays, in the design of communication systems the most important factor is usually the economical utilisation of the frequency spectrum even at the expense of complicating the design. If amplitude modulation is to be used, the SSB form of signal is preferred, since it requires the minimum bandwidth. For speech signals the separation of the upper and lower sidebands can be accomplished relatively easily, but for wideband applications the problems of filtering are more severe and can only be solved at considerable cost. A compromise solution, which is particularly successful for television transmission, uses a vestigial sideband (VSB) form of transmission consisting of a complete sideband plus a vestige of the other (see Fig. 2–7). Apart from the saving in cost due to the less stringent filter requirements compared with the formation of a SSB signal, there is also the advantage of being able to use an inexpensive demodulator—a feature of paramount importance in the design of a domestic receiver.

[2–1]

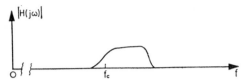

Fig. 2–7 Typical filter characteristic for the formation of a vestigial sideband signal.

2–2 Frequency and Phase Modulation

In the last section the amplitude of the carrier has been made to vary in sympathy with the modulating signal; the two parameters containing the message information, viz. the amplitude and rate (or frequency) at which it changes, determine the envelope pattern of the resultant signal. This, however, is not the only method of modulating a sinusoidal carrier. If we consider the general expression

$$e(t) = E_c\cos(\omega t + \phi) \qquad (2\text{–}9)$$

it can be seen that the frequency of the carrier, or its phase relative to some reference, is also available for carrying the information contained in the message.

Although frequency changes and phase changes are synonymous, there is a difference between what is known as *frequency modulation* (FM) and *phase modulation* (PM). Before embarking upon the mathematical analysis let us define the two forms to show how they are related to the modulating signal.

By definition, frequency modulation occurs when the carrier frequency is made to vary about its unmodulated value f_c in such a way that the instantaneous deviation is directly proportional to the instantaneous amplitude of the modulating signal. The rate at which the carrier frequency varies is identical to the frequency information of the message. Note that the carrier amplitude E_c is not included in this definition and may therefore be kept constant.

With phase modulation, the phase deviation of the carrier is directly related to the modulating signal amplitude and the rate at which it varies is identical to the frequency information of the message.

First consider the derivation of an expression to satisfy the definition of frequency modulation. The carrier frequency will vary from one instant to the next, or we may write the instantaneous frequency,

$$f = f_c + k_f m(t) \qquad (2\text{–}10)$$

where k_f is an arbitrarily chosen constant. For example, if the instantaneous amplitude of the modulating signal is 1 volt, the resultant carrier deviation can be, say, 1 kHz, 20 kHz or virtually any value—in practice there is a maximum limit and this is discussed later.

A phasor rotating with a constant angular velocity moves through a distance $\theta = \omega t$ in t seconds. This, of course, corresponds to the unmodulated carrier, i.e.

$$e(t) = E_c\cos\theta = E_c\cos\omega_c t \qquad (2\text{--}11)$$

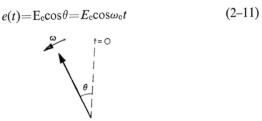

Fig. 2–8 Rotating phasor.

But when the carrier is frequency modulated the angular velocity is no longer constant and therefore

$$\theta = \int_o^t \omega dt$$

and using Eq. (2–10)

$$\theta = \int_o^t 2\pi[f_c + k_f m(t)]dt$$

$$= 2\pi f_c t + 2\pi\int_o^t k_f m(t)dt \qquad (2\text{--}12)$$

Therefore the expression for a frequency modulated signal becomes

$$e(t)\bigg|_{\text{FM}} = E_c\cos[2\pi f_c t + 2\pi\int_o^t k_f m(t)dt] \qquad (2\text{--}13)$$

Next consider the derivation of an expression for phase modulation. The instantaneous phase ϕ must be linearly related to the amplitude of the modulating signal, therefore

$$\phi = \phi_c + k_p m(t) \qquad (2\text{--}14)$$

where k_p is a constant similar to k_f in Eq. (2–10). Substituting Eq. (2–14) into Eq. (2–9) gives

$$e(t)\bigg|_{\text{PM}} = E_c\cos[\omega_c t + \phi_c + k_p m(t)]$$

and since ϕ_c is a constant it may be considered as zero; therefore

$$e(t)\bigg|_{\text{PM}} = E_c\cos[\omega_c t + k_p m(t)] \qquad (2\text{--}15)$$

[2–2]

Tutorial Exercise 2–1. What are the dimensions of k_t and k_p? The PM expression has been derived by substituting Eq. (2–14) into Eq. (2–9). Why could not a similar substitution of Eq. (2–10) in place of Eq. (2–14) have been used to establish the FM expression?

Eqs. (2–13) and (2–15) show that both FM and PM involve the variation of phase angle as a function of the modulating signal and because of this they are referred to under the general heading of *angle modulation*. Yet another definition refers to them as forms of *exponential modulation*, which can be appreciated if either Eq. (2–13) or Eq. (2–15) is written in exponential form. Taking Eq. (2–15),

$$e(t)\Big|_{PM} = \tfrac{1}{2} E_c\{\exp j[\omega_c t + k_p m(t)] + \exp j[-\omega_c t - k_p m(t)]\}$$

it is seen that the carrier is multiplied by the modulating function $\exp[jk_p m(t)]$, and hence the definition.

To allow a comparison of the expressions for FM and PM, and also to deduce the bandwidths of the signals, it is necessary to work with a specific type of modulating signal, since an analysis using the general function $m(t)$ is prohibitively difficult. The simplest function is $m(t) = E_m \cos\omega_m t$, i.e. a single modulating sinusoid. Substituting into Eq. (2–13) yields

$$e(t)\Big|_{FM} = E_c\cos[2\pi f_c t + 2\pi \int_0^t k_t E_m\cos\omega_m t \, dt] \qquad (2\text{–}16)$$

$$= E_c\cos\left[\omega_c t + \frac{k_t E_m}{f_m}\sin\omega_m t\right] \qquad (2\text{–}17)$$

The product $k_t E_m$ is the peak frequency deviation—this will be realised from the answer to the first part of the Tutorial Exercise above—and is given the symbol Δf. Therefore the maximum frequency excursion of the carrier is from $(f_c - \Delta f)$ to $(f_c + \Delta f)$.

The FM expression can also be written as

$$e(t)\Big|_{FM} = E_c\cos[\omega_c t + \beta\sin\omega_m t] \qquad (2\text{–}18)$$

where the *modulation index*, $\beta = \Delta f/f_m$ and is in radians.

The expression for PM with sinusoidal modulation is

$$e(t)\Big|_{PM} = E_c\cos[\omega_c t + k_p E_m\cos\omega_m t]$$

$$= E_c\cos[\omega_c t + \beta\cos\omega_m t] \qquad (2\text{–}19)$$

and in this equation $\beta = k_p E_m$. The reader may think it somewhat confusing to use the same symbol β for both FM and PM modulation indices. This, however, is common practice and surprisingly little confusion arises. Note the similarity of Eqs. (2–18) and (2–19). In fact, if a carrier is frequency or phase modulated by a single sine wave of constant peak amplitude E_m, an inspection of the modulated signal, say by some distant receiver, would not reveal whether the intended modulation was either frequency or phase. If, however, the frequency f_m of the modulating signal is changed, but the peak amplitude E_m kept constant, then for

(i) FM, the peak frequency deviation would remain the same, but the indirect phase modulation—due to $\Delta f/f_m$—would change, and for

(ii) PM, the peak phase displacement would remain the same but the indirect frequency modulation would change.

This is further illustrated by the phasor diagrams of Fig. 2–9. Consider the unmodulated phasor to be stationary on the paper and

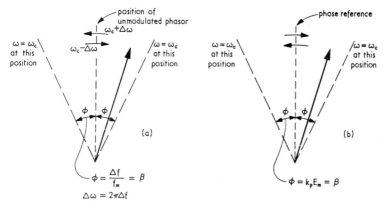

Fig. 2–9 Phasor diagrams for (*a*) frequency modulation, and (*b*) phase modulation, when the modulating signal is a sine wave.

vertically disposed. The modulating sinusoid will cause oscillation at a frequency f_m about this unmodulated position. For frequency modulation the peak deviation, i.e. the peak angular velocity, will be experienced as the phasor passes through the central position and is determined by the amplitude of the modulating signal ($\Delta f = k_f E_m$). The peak phase displacement is equal to $\Delta f/f_m$—note the smaller f_m the larger the phase displacement. For phase modulation the peak phase displacement is fixed by the modulating signal amplitude E_m and is independent of f_m, but the angular velocity in the central position (i.e. the unavoidable frequency modulation) is directly proportional to f_m.

Having defined the two forms of angle modulation and derived expressions for the resultant signals when the modulation is due to a

[2–2]

single sinusoid, we must next determine their bandwidths. The AM signal analysis presented little difficulty, since the sidebands were deduced from a simple trigonometrical expansion. Angle modulation presents a more complicated problem. An expansion of Eq. (2–18) gives

$$e(t)\Big|_{\text{FM}} = E_c[\cos\omega_c t \cos(\beta\sin\omega_m t) - \sin\omega_c t \sin(\beta\sin\omega_m t)] \quad (2\text{--}20)$$

which requires further manipulation before the sideband structure can be determined. The terms cos(sin) and sin(sin) each represent a power series of a power series. The study of these functions was originally the work of F. W. Bessel (1784–1846), the German astronomer, and they are now known by his name. For our present requirements it is not necessary to know about the calculus of Bessel Functions and we shall simply use the notations as a convenient form of shorthand, i.e. the terms may be expressed as

$$\cos(\beta\sin\omega_m t) = J_0(\beta) + 2\sum_{n=1}^{\infty} J_{2n}(\beta)\cos(2n\omega_m t)$$

and

$$\sin(\beta\sin\omega_m t) = 2\sum_{n=0}^{\infty} J_{2n+1}(\beta)\sin(2n+1)\omega_m t$$

where

$$J_n(\beta) = \sum_{m=0}^{\infty} \frac{(-1)^m(\beta/2)^{2m+n}}{m!(m+n)!}$$

and $J_0(\beta)$ is known as the zero-order Bessel Function, $J_1(\beta)$ the first-order Bessel Function and so on. Using these expressions in Eq. (2–20) gives

$$e(t)\Big|_{\text{FM}} = E_c\Big\{J_0(\beta)\cos\omega_c t + \sum_{n=1}^{\infty} J_{2n}(\beta)[\cos(\omega_c+2n\omega_m)t + \cos(\omega_c-2n\omega_m)t]$$

$$- \sum_{n=0}^{\infty} J_{2n+1}(\beta)[\cos(\omega_c-\overline{2n+1}\omega_m)t - \cos(\omega_c+\overline{2n+1}\omega_m)t]\Big\}$$

and rearranging, gives

$$e(t)\Big|_{\text{FM}} =: E_c\Big\{J_0(\beta)\cos\omega_c t$$

$$+ J_1(\beta)[\cos(\omega_c+\omega_m)t - \cos(\omega_c-\omega_m)t]$$

$$+ J_2(\beta)[\cos(\omega_c+2\omega_m)t + \cos(\omega_c-2\omega_m)t]$$

$$+ J_3(\beta)[\cos(\omega_c+3\omega_m)t - \cos(\omega_c-3\omega_m)t]$$

$$+ \ldots \ldots\Big\} \quad (2\text{--}21)$$

A similar expression can be obtained for the phase-modulated signal. Eq. (2–21) is a startling result, for we see that there are an infinite number of pairs of sidebands* harmonically related to the modulating frequency f_m, which implies that the bandwidth of the signal is infinite. The amplitude of the carrier depends upon the coefficient $J_0(\beta)$ and the sideband amplitudes according to $J_1(\beta)$, $J_2(\beta)$, etc. Let us use the tables of Appendix IV, page 282, and the graphs of Fig. 2–10 to compare their amplitudes for various values of β: remember that β can be arbitrarily chosen because it involves the constant k_f.

First, when $\beta \ll 1$, $J_0(\beta) \rightarrow 1$ and $J_1(\beta)$, $J_2(\beta)$, etc.$\rightarrow 0$; hence the sidebands are negligible and only the carrier is present in the output. This is of no value, since the information content is zero. As the frequency

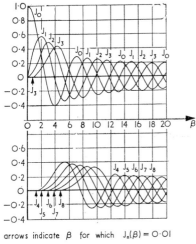

arrows indicate β for which $J_n(\beta) = 0.01$

Fig. 2–10 Bessel functions.

deviation is increased by increasing the amplitude of the modulating sinusoid—and assuming its frequency remains constant—the modulation index is made larger and consequently sidebands having significant amplitudes appear in the output. Examples of the FM spectra for β between 0·4 and 8 are shown in Fig. 2–11.

In practice there is always a limited bandwidth available for a particular transmission and therefore a compromise must be reached when choosing β. On the one hand β should be made as large as possible for maximum enhancement of the information, but contrary to this requirement it should be made sufficiently small to avoid the generation of side-

* The modulation process is said to be *non-linear* because there is not a one-to-one relationship between input and output components.

[2–2]

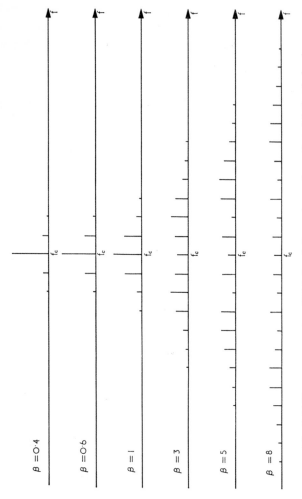

Fig. 2–11 Frequency spectra of FM signals for various values of modulation index β. (Only the significant sidebands are shown.)

bands that would cause interference to adjacent channels. Theoretically the FM signal occupies an infinite bandwidth and it is impossible not to generate interfering sidebands. However, these may be considered insignificant if their amplitudes are less than 1% of the unmodulated carrier. This is an arbitrary but generally accepted criterion and means that the nth pair of sidebands are insignificant when

$$J_n(\beta) < 0.01 J_0(0)$$

where $J_0(0)$ is the coefficient of the unmodulated carrier. But since $J_0(0)=1$, we have the significant sideband criterion defined by

$$J_n(\beta) > 0.01$$

Care must be exercised when using this criterion; for example, when $\beta = 7$ the first pair of sidebands have coefficients of $J_1(7)=0.47\%$ of the unmodulated carrier. Although this pair is less than 1%, the higher orders $J_2(7)$, $J_3(7)$ are indeed significant. In fact, for this particular value of β the sidebands become insignificant after the tenth pair (see Table IV–2, page 284). A selection of modulation indices with the corresponding significant sidebands are listed in Table 2–1.

Table 2–1

Modulation index, β	No. of *pairs* of significant sidebands
0·2	1
0·4	2
0·6	2
0·8	3
1·0	3
1·5	4
3·0	6
5·0	8
8·0	11
10·0	14
20·0	25

The determination of the significant bandwidth of an angle-modulated signal has been carried out for single sine-wave modulation. Unfortunately, a simple extension of the analysis to accommodate a complex baseband signal, as for amplitude modulation, does not follow because of the lack of a one-to-one correspondence between input and output components and the consequent violation of the theorem of Superposition. Nevertheless, it is possible to get some idea of the modulated signal bandwidth by considering the sidebands due to a component at

[2–2]

the highest frequency in the baseband. Take, for example, the transmission of high-quality sound occupying a frequency range from 50 Hz–15 kHz. Let us assume that a channel with a 250 kHz bandwidth is available in the VHF band; the question is, 'What value of modulation index β should be chosen to make optimum use of the allocated channel and to cause minimum interference to adjacent channels when using frequency modulation?' Consider the highest frequency in the baseband, i.e. 15 kHz. The eighth pair of sidebands in the FM signal would be at $f_c \pm (8 \times 15) = f_c \pm 120$ kHz, and these must be considered significant for transmission, since they fall within the channel allocation of $f_c \pm 125$ kHz. The ninth pair of sidebands will be outside this band and therefore β must be chosen to make $J_9(\beta)$ and all higher order coefficients less than 0·01. From Table 2–1 or Table IV–2, page 284, we see that $\beta = 5$ will satisfy this requirement. Using the relation $\Delta f = \beta f_m$, the peak frequency deviation of the carrier is $\pm (5 \times 15) = \pm 75$ kHz.

Tutorial Exercise 2–2. Continue with the above example and check that the sidebands due to lower frequency baseband components are insignificant outside the band $f_c \pm 125$ kHz. Assume that the peak amplitude of any component in the range 50 Hz–15 kHz can produce a carrier deviation of ± 75 kHz.

It is to be noted in the above example that a peak frequency deviation of the carrier of ± 75 kHz (i.e, a total of 150 kHz) results in a signal that occupies a significant bandwidth of 240 kHz. The reader will also realise from a study of the Tutorial Exercise that the lower the modulating frequency the greater the number of sidebands in the output. It has been arranged that 15 kHz modulation gives rise to 8 pairs of significant sidebands and it can be shown—as part of the above exercise—that when the modulating frequency is, say, 6 kHz there are 16 pairs of significant sidebands contained in a total bandwidth of 192 kHz. For phase modulation the amplitudes of the sidebands are determined by the Bessel coefficients $J_n(\beta)$, as for FM, but $\beta = k_p E_m$ and not $\Delta f/f_m$. Using the figures of the above example, a modulation index $\beta = 5$ is necessary to satisfy the significant sideband criterion when the modulating frequency f_m is 15 kHz, but since β is independent of f_m there can never be more than 8 pairs of significant sidebands corresponding to any component in the baseband range 50 Hz–15 kHz. We see, therefore, that a 6 kHz modulation frequency would result in a PM signal having a spectrum $f_c \pm 48$ kHz, and at the extreme of a 50 Hz modulating signal the occupied bandwidth would be only 800 Hz. We may conclude that phase modulation does not make efficient use of the bandwidth available. In fact, this is one of the reasons why FM is preferred for the transmission of analogue signals; another is that the receiver must have a reference phase when demodulating a PM signal, and this is usually difficult to achieve with the required degree of accuracy.

A special case of analogue modulation arises when $\beta \ll 1$, and to distinguish it from the general case (known as *wide-band* FM or PM) where β can have any value, it is referred to as *narrow-band* modulation. Consider Eq. (2–20), i.e.

$$e(t)\Big|_{\text{FM}} = E_c[\cos\omega_c t \cos(\beta\sin\omega_m t) - \sin\omega_c t \sin(\beta\sin\omega_m t)]$$

When β is the order of 0·2 radian or less, then

$$\cos(\beta\sin\omega_m t) \approx 1 \text{ and } \sin(\beta\sin\omega_m t) \approx \beta\sin\omega_m t$$

and the FM signal is given by

$$e(t)\Bigg|_{\substack{\text{FM} \\ \text{narrow-band}}} \approx E_c[\cos\omega_c t - \beta\sin\omega_m t\sin\omega_c t] \tag{2–22}$$

$$= E_c\left\{\cos\omega_c t - \frac{\beta}{2}\cos(\omega_c - \omega_m)t + \frac{\beta}{2}\cos(\omega_c + \omega_m)t\right\} \tag{2–23}$$

It is interesting to compare Eq. (2–22) with Eq. (2–4) representing the full AM signal,

$$e(t)\Bigg|_{\substack{\text{full} \\ \text{AM}}} = E_c[\cos\omega_c t + m_a\cos\omega_m t\cos\omega_c t] \tag{2–24}$$

$$= E_c\left[\cos\omega_c t + \frac{m_a}{2}\cos(\omega_c - \omega_m)t + \frac{m_a}{2}\cos(\omega_c + \omega_m)t\right] \tag{2–25}$$

We see that both forms are made up of a carrier and two sidebands, but for AM the sidebands are associated with an in-phase carrier $\cos\omega_c t$, and for narrow-band FM, with a quadrature-phase carrier $\sin\omega_c t$, compare Eq. (2–24) with Eq. (2–22). This can be further illustrated by representing Eqs. (2–23) and (2–25) in exponential form to permit the establishment of phasor diagrams. We have

$$e(t)\Bigg|_{\substack{\text{FM} \\ \text{narrow-band}}} = E_c\text{Re}\left\{\exp(j\omega_c t)\left[1 - \frac{\beta}{2}\exp(-j\omega_m t) + \frac{\beta}{2}\exp(j\omega_m t)\right]\right\} \tag{2–26}$$

and

$$e(t)\Bigg|_{\substack{\text{full} \\ \text{AM}}} = E_c\text{Re}\left\{\exp(j\omega_c t)\left[1 + \frac{m_a}{2}\exp(-j\omega_m t) + \frac{m_a}{2}\exp(j\omega_m t)\right]\right\} \tag{2–27}$$

Re signifies that $e(t)$ is real at any instant in time. The complex form of the components inside the curly brackets gives the instantaneous phase relationships.

It is left as an exercise for the student to show that

$$e(t)\bigg|_{\substack{\text{PM}\\ \text{narrow-}\\ \text{band}}} = E_c\text{Re}\left\{\exp(j\omega_c t)\left[1-j\frac{\beta}{2}\exp(-j\omega_m t)-j\frac{\beta}{2}\exp(j\omega_m t)\right]\right\} \quad (2\text{--}28)$$

The phasor diagram of Fig. 2–12(a) indicates that the sidebands of an AM signal are always symmetrically disposed about the carrier and therefore the resultant signal varies in amplitude but not in frequency— as required. Narrow-band FM and PM have sidebands symmetrically positioned about a quadrature carrier position and each gives rise to a resultant signal whose phase displacement is a function of the modulating signal. Theoretically, the peak amplitude of an angle-modulated signal is only constant when there are an infinite number of sidebands present. The assumptions made for narrow-band modulation, leading up to Eq. (2–22), have produced a signal with simply a carrier and the

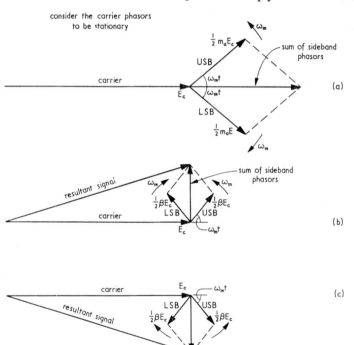

Fig. 2–12 Phasor diagrams: (a) full amplitude modulation; (b) narrow-band frequency modulation, and (c) narrow-band phase modulation.

first pair of sidebands, so it is not surprising to see from Figs. 2–12(*b*) and (*c*) that the angle-modulated signals have a small amount of amplitude modulation as well. The application of this form of modulation is dealt with in the next chapter.

At this stage the reader may wonder why angle modulation is considered at all, since its wideband requirement, compared with AM, contradicts the comments made in Chapter 1 regarding the need to economise on bandwidth. The answer to such a query must be left until we are able to discuss the reception of signals in the presence of noise (see Chapter 5).

Further Tutorial Exercises

2–3 Two transmitters, one capable of generating a full AM signal and the other a SSB signal, have equal mean output power ratings. A single sine wave modulating signal causes the output from the full AM transmitter to have a modulation index of 0·8. If both transmitters are operating at maximum output, compare, in dB, the power contained in the sidebands of the full AM signal to that contained in the SSB output.

Ans: power ratio, AM sidebands/SSB $\equiv -6\cdot2$ dB.

2–4 The instantaneous amplitude of a full AM signal is given by the expression

$$e(t) = 1000\left[1 + \sum_{n=1}^{3} \frac{1}{2n}\cos(2\pi n \times 10^3 t)\right]\cos(2\pi \times 10^7 t) \text{ volts}$$

Determine the peak amplitudes and frequencies of the various components of the modulated signal and plot the amplitude-frequency spectrum. Calculate the mean power that the signal would dissipate in a 50-ohm resistive load.

Ans: 1000 V, 10^7 Hz;
250 V, $(10^7 \pm 10^3)$ Hz;
125 V, $(10^7 \pm 2 \times 10^3)$ Hz;
83 V, $(10^7 \pm 3 \times 10^3)$ Hz;
11·7 kW.

2–5 A bandwidth of 25 kHz is allocated in the UHF spectrum for the transmission of a single speech channel. Using the significant sideband criterion stated on page 39, estimate the peak frequency deviation of a suitable frequency modulated signal carrying the speech information. Values of Bessel Functions can be obtained from Appendix IV.

Ans: $\pm$ 4·8 kHz when the highest speech frequency is 3·4 kHz;
$\pm$ 6·3 kHz when the highest speech frequency is 3 kHz.

2–6 In a frequency modulation system the allocated channel bandwidth is 11 MHz and the total transmitter power is 10 W. It is specified that there shall be no components outside the channel band with power levels exceeding 10 mW. Estimate, to within 1%, the maximum peak frequency deviation that can be used with single tone modulation at 1 MHz. Values of Bessel Functions can be obtained from Appendix IV.

Ans: 3·65 MHz.

3
GENERATION AND RECEPTION OF MODULATED SIGNALS

The aim of this chapter is to outline in general terms the more common methods of generating the various forms of amplitude and angle modulation; the inverse procedure, known as demodulation, for the recovery of the baseband information is considered for noise-free reception. The presence of noise and its effect upon the choice of modulation system and method of demodulation is dealt with in Chapter 5.

3-1 Amplitude Modulation

It was shown in the previous chapter that the essential requirement is the multiplication of the baseband signal with a higher frequency sinusoid. This seemingly straightforward operation presents considerable difficulty in practice, since there are relatively few devices that can be called multipliers; probably the best known is the Hall Effect Multiplier, but this has not received much attention for communication purposes, due mainly to its high cost compared with other methods, to be described, and its low conversion gain. There are, however, many devices that possess a multiplying capability as just one of the many features of a complicated non-linear characteristic. For our application this would mean an output of the form $m(t)[\cos\omega_c t]$ accompanied by higher-order multiplicative terms. There are, in the main, two solutions to this problem. One is to follow a 'non-perfect' multiplier with a filter that allows only the wanted components to pass, whilst the other is to arrange a pair of non-linear elements, having as near as possible the same characteristics, in a balanced circuit so that the unwanted components cancel out. The degree of suppression depends upon the closeness of matching, and in practice it is usually found necessary to provide further suppression by adding an output filter—this requirement, however, need not be as stringent as that for the first method.

Before a comparison of the two methods can be made, it is necessary to consider a typical non-linear device such as a semiconductor diode, and to establish its input/output relationships for conditions of small and large signal operations. The analyses of the following sections deal exclusively with the diode, since both of the above-mentioned possibilities can be realised and, furthermore, the fundamental ideas can be readily applied to modulators using bipolar transistors and MOSFETs.

3–1.1 *Small-Signal Modulators*

When a diode is operated over a restricted portion of its characteristic (see Fig. 3–1) the current $i(t)$, and hence the voltage $v_{out}(t)$ across a resistive load (see Fig. 3–2 (*a*)), is related to the input voltage $v(t)$ by the polynomial

$$v_{out}(t) = i(t)R = av(t)+bv^2(t)+cv^3(t) \qquad (3\text{--}1)$$

The coefficients a, b and c may be assumed constant and higher-order terms sufficiently small to be neglected. Consider the input to be made up of a modulating voltage $m(t)$ and a sine wave $E_c\cos\omega_c t$. The output is

$$
\begin{aligned}
v_{out}(t) = {} & a[E_c\cos\omega_c t+m(t)] \\
& +b[E_c\cos\omega_c t+m(t)]^2 \\
& +c[E_c\cos\omega_c t+m(t)]^3 \\
= {} & aE_c\cos\omega_c t+am(t) \\
& +bm^2(t)+2bm(t)E_c\cos\omega_c t+bE_c^2\cos^2\omega_c t \\
& +cm^3(t)+3cm^2(t)E_c\cos\omega_c t+3cm(t)E_c^2\cos^2\omega_c t+cE_c^3\cos^3\omega_c t
\end{aligned}
$$
$$(3\text{--}2)$$

We see that the term giving rise to the required upper and lower sidebands is $2bm(t)E_c\cos\omega_c t$. All the other terms produce outputs harmonically related to the modulating signal, carrier and modulated carrier. For example, those of the form $m(t)E_c^2\cos^2\omega_c t$ are responsible for sidebands at twice the carrier frequency and although undesirable it is relatively simple to remove them by filtering. The same cannot be said of the outputs due to the term $cm^2(t)E_c\cos\omega_c t$. For the purpose of illustration let us assume that

$$m(t) = \cos\omega_1 t+\cos\omega_2 t \qquad (3\text{--}3)$$

then $cm^2(t)E_c\cos\omega_c t=c(\cos^2\omega_1 t+2\cos\omega_1 t\cos\omega_2 t+\cos^2\omega_2 t)E_c\cos\omega_c t$ and an expansion yields components at frequencies f_c, $f_c\pm2f_1$, $f_c\pm2f_2$, $f_c\pm(f_1+f_2)$ and $f_c\pm(f_1-f_2)$. It will be appreciated that these can fall within the bandwidth occupied by the required DSB output and therefore cannot be removed by filtering. We conclude that the single diode modulator is not a practical proposition, since these components, due to *intermodulation* within the baseband signal, would cause interference to the information contained in the modulated output.

$$[3\text{--}1]$$

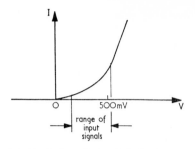

Fig. 3–1 Typical small-signal diode characteristic.

By using two matched diodes in a circuit that is symmetrical (or balanced) with respect to the carrier input (see Fig. 3–2(b)), the unwanted intermodulation components can be cancelled out. In order to

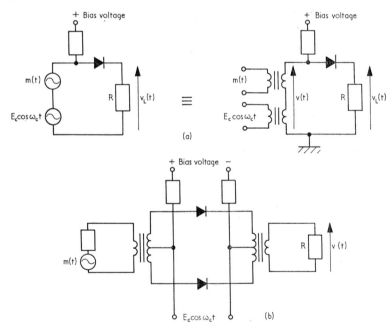

Fig. 3–2 (a) Single-diode modulator, and (b) a balanced modulator.

facilitate the explanation of the balanced modulator a transformer arrangement is used, although in practice it is often more convenient to employ active devices. The modulator output voltage is

$$v_{\text{out}}(t) = \text{constant} \times [i_1(t) - i_2(t)] \qquad (3\text{–}4)$$

where $i_1(t)$ is the current through diode 1 due to the applied voltage

$[E_c\cos\omega_c t + m(t)]$ and $i_2(t)$ is the current through diode 2 due to the applied voltage $[E_c\cos\omega_c t - m(t)]$. Substituting these two conditions into Eq. (3–1) and then using Eq. (3–4) gives

$$v_{out}(t) \propto 2am(t) + 4bm(t)E_c\cos\omega_c t$$
$$+ 6cm(t)E_c{}^2\cos^2\omega_c t + 2cm^3(t) \qquad (3\text{–}5)$$

It is to be noted from the above equation that the small-signal balanced modulator has produced a DSBSC output together with unwanted terms that can be filtered out.

3–1.2 *Large-Signal* Modulators (Piecewise-Linear Modulators, Chopper Modulators)*

It is also possible to obtain a modulated output using the circuit of Fig. 3–2(*b*), by operating the diodes in a different way from that described above. Instead of a low-level carrier (peak value of hundreds of millivolts) a relatively high level of 3 or 4 volts can be used to switch the diodes from a reversed biased condition to one of forward conduction—the d.c. biasing no longer being required. If the diodes are driven well into the forward biased region, each will have a current-voltage relationship that is essentially linear, due to the diode acting more like a resistor than a semiconductor. Under these large-signal conditions the overall characteristic is made up of two linear portions (see Fig. 3–3),

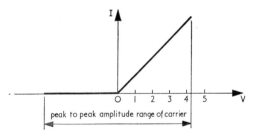

Fig. 3–3 Typical large-signal diode characteristic.

and hence we have the explanation of the term *piecewise linear* modulator. The switching of the diodes causes the low-level modulating voltage to appear across the output every other half-cycle of the carrier† and therefore

---

* It must be understood that the definition *large-signal* is only used to distinguish this mode of operation from the previous small-signal case: it does not mean high-power.

† The output is a chopped version of the input—hence the description 'Chopper Modulator'.

[3–1]

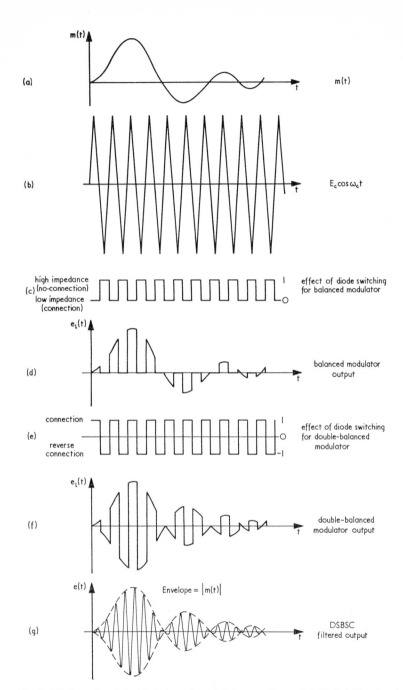

Fig. 3–4 Balanced and double-balanced modulator waveforms. (*a*) Modulating signal $m(t)$. (*b*) Large-signal carrier $E_c\cos\omega_c t$. (*c*) Switching action of a balanced modulator. (*d*) Output from a balanced modulator. (*e*) Switching action of a double-balanced modulator. (*f*) Output from a double-balanced modulator. (*g*) DSBSC filtered output.

$$v_{out}(t) = m(t) \times \begin{bmatrix} squared\ waveform\ due\ to\ switching \\ of\ the\ diodes\ by\ the\ carrier \end{bmatrix}$$

i.e. $v_{out}(t) \propto m(t)\left[\frac{1}{2}+\frac{2}{\pi}(\cos\omega_c t-\frac{1}{3}\cos3\omega_c t+\frac{1}{5}\cos5\omega_c t+ \ldots)\right]$

$$= \frac{1}{2}m(t)+\frac{2}{\pi}m(t)\cos\omega_c t+ \text{ terms involving } 3f_c, 5f_c, \text{ etc. } \quad (3\text{--}6)$$

The bracketed expression is the Fourier series representation of the switching action. We see that the term $2[m(t)\cos\omega_c t]/\pi$ will provide a DSBSC output and that it is independent of the peak amplitude of the carrier E_c. A typical modulated output is shown in Fig. 3–4 (*d*).

Another version known as a *double-balanced* modulator has four diodes—the additional two connected in a lattice arrangement as shown in Fig. 3–5. The switching action of the carrier causes the modulating input $m(t)$ to appear across the output with what is effectively a reversal of the connections every other half-cycle of the carrier. The switching waveform is illustrated in Fig. 3–4(*e*) and the output is

$$v_{out}(t) = m(t) \times \begin{bmatrix} squared\ waveform\ corresponding \\ to\ switching\ reversals \end{bmatrix}$$

$$= m(t)\left[\frac{4}{\pi}(\cos\omega_c t+\frac{1}{3}\cos3\omega_c t+\frac{1}{5}\cos5\omega_c t+ \ldots)\right]$$

$$= \frac{4}{\pi}m(t)\cos\omega_c t+ \text{ terms involving } 3f_c, 5f_c, \text{ etc. } \quad (3\text{--}7)$$

It is seen that the only difference between this output and that of the balanced modulator, given by Eq. (3–6), is a component at the modulating frequency $m(t)$.

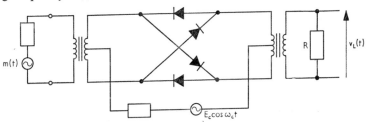

Fig. 3–5 Double-balanced modulator.

3–1.3 *Comparison of Small-Signal and Large-Signal Modulators*

After reading the previous two sections, the large-signal modulator would appear to be the obvious choice for generating a DSBSC signal. Before dismissing the small-signal modulator completely it is worth

[3–1]

while to state firstly the conditions under which it can be used and secondly to enumerate the advantages of the large-signal circuit. The latter becomes the obvious choice when diodes of a discrete form are used, as it is not practicable to select pairs having matched coefficients *a*, *b* and *c*—a necessary requirement to avoid intermodulation when using a small-signal modulator. The matching of diodes for large-signal application presents little problem, since the forward bias *I-V* characteristic does not differ appreciably from diode to diode within a production batch, and the large discrepancies in reverse bias characteristics can be overcome by shunting each diode with a high resistance, say 100 kΩ.

Balanced modulators in the form of integrated circuits using bipolar transistors in place of diodes—simply for convenience of manufacture— can be made up of pairs of elements having characteristics that would be considered matched if exhibited by a pair of discrete devices. Nevertheless, these circuits are used solely as large-signal modulators mainly due to the convenience of not having to supply a d.c. bias and also, of course, of making use of the overriding advantage that spurious outputs due to intermodulation are extremely small even when the non-linear elements are not identical. But integrated circuits made up of unipolar transistors (FETs and MOSTs) do, however, lend themselves to small-signal modulation. Both of these transistors exhibit an extremely good square-law characteristic, i.e. the $bv^2(t)$ term of Eq. (3–1) is several orders of magnitude greater than the other terms and, therefore, intermodulation outputs are virtually non-existent. Further reference is made to these devices in the section dealing with frequency changers and demodulators.

3–1.4 *High-Power Modulators*

A full AM signal can be generated by operating an amplifier in a non-linear mode. The term *high-power* is used to indicate that the modulated signal can be formed at the required power for transmission and that no further amplification or signal processing is necessary. This method has been used extensively in mobile systems having outputs from tens to

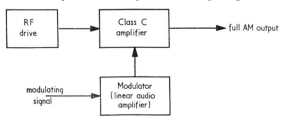

Fig. 3–6 High-power generation of a full AM signal.

hundreds of watts and also for broadcast and point-to-point radio links with outputs of many kilowatts.

The output stage of the transmitter is used as the modulator and consists of a thermionic valve (or valves) biased to operate as a Class C amplifier. Its output power is made to vary in sympathy with the modulating input derived from a linear audio amplifier (see Fig. 3–6). The main attraction of this arrangement is its extreme simplicity compared with other methods of generating a full AM signal, and whilst its use for mobile purposes can be tolerated it is not recommended for the generation of very high powers, because the spurious modulator outputs —those at $2f_c$, $3f_c$, etc.—cannot be filtered out and may cause considerable interference to other users of the radio spectrum. Alternative schemes are considered in section 3–2.

A full account of Class C amplifiers and high-power modulators is well documented elsewhere.[1, 2]

3–1.5 *SSB Generation*

The obvious way to generate a SSB signal is to pass the DSB output from a balanced modulator through a filter that rejects one of the side-

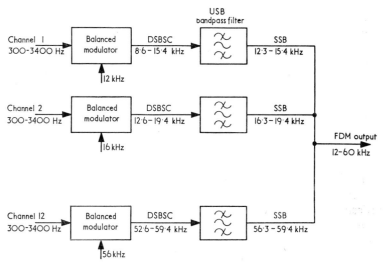

Fig. 3–7 Formation of FDM basic group.

1. *Electronic and Radio Engineering*, F. E. Terman, Chapter 13 (McGraw-Hill, 1955).

2. A Non-Linear Theory of Class C Transistor Amplifiers and Frequency Multipliers, R. G. Harrison, *Trans. I.E.E.E.*, Vol. SC-2, No. 3, pp. 93–102 and p. 126, September 1967.

[3–1]

bands. Its application for the generation of a FDM Basic Group for line communication is illustrated in Fig. 3–7. The outputs from the balanced modulators are filtered to remove the lower sideband of each channel.

It is also common practice to use this method for radio systems. However, the removal of a sideband that is closely spaced to the wanted sideband presents insuperable difficulties at very high carrier frequencies; the separation can only be successfully achieved at a relatively low carrier frequency f_{c_1}, of less than 2MHz. The technique is illustrated in Fig. 3–8, and it is also shown how two *independent sidebands* (ISB), symmetrically positioned about the carrier, can be formed. It is basically an SSB system, but has the economic advantage of allowing one piece of equipment to handle two messages. The ISB form of signal is used mainly for the transmission of telephony in the HF (3–30 MHz) band. The propagation characteristics of this part of the spectrum, together with the limited number of channels, preclude the use of wideband systems carrying a large number of messages, whereas the use of ISB to provide two (or four) channels on the same link is feasible and economically desirable [3]

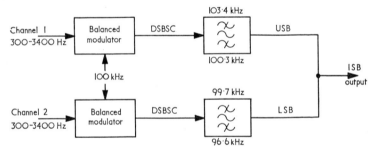

Fig. 3–8 Formation of an ISB signal.

Having formed the SSB or ISB output at a relatively low carrier frequency and low power, usually of the order of 100 mW–1 W, further processing is needed to transpose the signal to a higher frequency and to amplify it to the required output level. It follows that the frequency translation must involve another process of modulation where the modulating signal is the SSB version of the original baseband input $m(t)$. This modulator (or frequency changer) will produce a DSB output symmetrically disposed about the new carrier frequency f_{c_2} (see Fig. 3–9). The frequency separation of the sidebands is twice the carrier frequency f_{c_1}, which therefore facilitates the removal of the unwanted

    3. *High Frequency Communications*, J. A. Betts, Chapters 3 and 5 (English Universities Press, 1967).

sideband. If the final output is required in the VHF or UHF spectrum, yet another stage of frequency changing is needed—the number of transpositions being determined solely by the final carrier frequency and

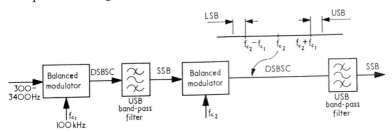

Fig. 3–9 Generation of an SSB signal by successive modulation and filtering.

the ability of the filters to separate the upper and lower sidebands after each stage of modulation.

A phase-shift method of generating an SSB signal[4, 5] (see Fig. 3–10) does not require filters having sharp cut-off frequency characteristics. Two balanced modulators generate DSB signals that, when added together, result in the cancellation of either the lower or upper sidebands.

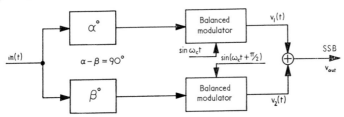

Fig. 3–10 Phase-shift method of generating a SSB signal.

This condition is realised when the modulating inputs and the carriers, each differ by 90°. The output from the quadrature-phase modulator, as it is known, is

$$v_{out}(t) = v_1(t) + v_2(t) \tag{3-8}$$

where $v_1(t) = E_m\cos\omega_m t\sin\omega_c t$

and $v_2(t) = E_m\cos(\omega_m t - \pi/2)\sin(\omega_c t + \pi/2)$

The modulating input $m(t)$ is assumed to be a single sine wave $E_m\cos\omega_m t$

4. The Phase-Shift Method of Single Sideband Signal Generation, D. E. Norgaard, *Proc. I.R.E.*, Vol. 44, No. 12, pp. 1718–35, December 1956.

5. A New Version of the Out-phasing Method for Frequency Translation (SSB Generation and Detection), W. Saraga, *Transmission Aspects of Communication Networks*, *I.E.E. Conf. Publ.* ED7, pp. 131–4, February 1964.

[3–1]

—a justifiable simplification for the linear modulation process, see page 25. Substituting for $v_1(t)$ and $v_2(t)$ in Eq. (3–8) gives

$$v_{out}(t) = m(t)\sin(\omega_c + \omega_m)t$$

The main difficulty in practice is to achieve a constant 90° phase shift over the spectrum of the modulating input. For a speech input it can be realised by having phase-shift networks in each path with their difference $(a - \beta)$, see Fig. 3–10, maintained at a constant value by differential control: further details can be found in Reference 4.

Tutorial Exercise 3–1. A third method of generating a SSB signal, due to D. K. Weaver, is described in Reference 6. Examine the frequency spectra of the outputs of the first pair of balanced modulators. Insert baseband frequencies of 300–3400 Hz, i.e. representing a speech input, into the general expressions and thence calculate the spectrum of the SSB output when the second pair of modulators is supplied with a 1·2 MHz carrier.

3–2 Detection (Demodulation) of AM Signals not accompanied by noise

The demodulation of a full AM signal, i.e. DSB with carrier, differs considerably from that of a DSBSC or an SSB signal. It can be seen from Fig. 2–5 that the message $m(t)$ is contained in the envelope of the AM signal and therefore the detector can be a rectifier that responds to the peak amplitude of the input. This is referred to as *non-coherent* detection, since the demodulator does not require any information about the incoming signal. The envelope of a DSBSC or SSB signal is not a replica of $m(t)$ and, consequently, a simple rectifier will not serve as a suitable detector. It is shown in section 3–2.3 that for suppressed carrier modulation there is insufficient information for a non-coherent detector to recover the baseband information without distortion. The carrier must be inserted, at the receiver, before demodulation can be effected—a condition that defines a *coherent* detector.

A distinction can be made between the reception of a signal transmitted via a line (or cable) and one by radio. For the latter the received signal power is extremely small ranging from 10^{-4} to 10^{-6} W for short-range broadcast reception to 10^{-10} to 10^{-14} W received from a satellite transmitter. Furthermore, it is accompanied by a host of unwanted signals and general background interference. Before demodulation, the wanted signal must be extracted by filtering and, since considerable amplification is also required before the final output can operate a transducer, the filtering and gain are achieved in the same process, namely that of tuned amplification. The problems of line communication are of

6. A Third Method of Generation and Detection of Single Sideband Signals, D. K. Weaver, *Proc. I.R.E.*, Vol. 44, No. 12, pp. 1703–5, December 1956.

a completely different nature. The received signal power is considerably higher, due, of course, to successive amplification by repeaters spaced along the line. Whilst it is not intended to deal with repeater systems, the reader should not gain the impression that line communication is relatively straightforward—perhaps it should be mentioned that an ocean cable system has repeaters providing an overall gain of many thousands of decibels which must be stabilised to within 1 dB.[7]

Before dealing with the different methods of demodulation let us consider the prerequisites of amplification and filtering. A number of high-Q tuned amplifiers in tandem could provide the gain and selectivity if the receiver were for use at a fixed frequency, but for a variable tuning facility the various stages cannot be matched (or ganged) sufficiently closely for the required amplification to be maintained over the tuning range of the receiver. We see, therefore, that high selectivity and high sensitivity are incompatible with this form of receiver. An alternative is to use an amplifier having a high gain over a fixed range of frequencies, i.e. a band-pass amplifier, preceded by a frequency changer that transposes the incoming signal in a linear manner so that it is within the pass-band of the amplifier. This is known as a superheterodyne receiver. We shall not consider a detailed study of its design, since circuit techniques are largely determined by the particular frequencies of the incoming signals. The fundamental principles, however, are common to all receivers whether they be for VLF or SHF reception.

It was discussed in Chapter 2 that amplitude modulation can be regarded as a linear process due to the one-to-one correspondence of the input and output components. The frequency translation which must precede the band-pass amplifier—or *intermediate-frequency* (IF) amplifier as it is known, because the signal is neither at the original input frequency nor at the final baseband frequency—can therefore be achieved by a modulator that multiplies the incoming signal with a locally generated sine wave $E_0\cos\omega_0 t$. For the purpose of illustration let us assume that the input frequency is f_1—we shall ignore its bandwidth on the assumption that it is extremely small compared with f_1. The output of the frequency changer will consist of upper and lower sidebands, $f_1 \pm f_0$, and higher-order components involving $2f_0$, $3f_0$, etc. The intermediate-frequency amplifier is designed to allow one of these sidebands to experience full gain and to attenuate effectively the other sideband and higher-order components. Such a receiver is capable of amplifying and selecting incoming signals over a range of frequencies by

7. Anglo-Canadian Transatlantic Telephone Cable (CANTAT), J. F. Brampton et al., *Proc. I.E.E.*, Vol. 110, No. 7, pp. 1115–23, July 1963.

having a variable-frequency local oscillator f_0 that can be adjusted to satisfy

$$f_1 - f_0 = f_{IF} \qquad (3\text{-}9)$$

when f_{IF}, the centre frequency of the IF amplifier passband, is lower than the incoming signal frequency; or

$$f_1 + f_0 = f_{IF} \qquad (3\text{-}10)$$

if the receiver is designed with an IF that is higher than the input frequency. Referring to Fig. 3–11 and assuming that the receiver is designed for Eq. (3–9) to apply, the outputs from the frequency changer corresponding to inputs at f_1 and f_2 are $f_1 \pm f_0$ and $f_0 \pm f_2$ respectively, where $f_1 > f_0 > f_2$. If the two input frequencies are symmetrically positioned with respect to f_0, i.e. $f_1 - f_0 = f_0 - f_2$, the IF amplifier will be unable to distinguish between them and consequently serious mutual interference would result. The effect of the unwanted input (or *second channel*) can be reduced by having a high IF—large separation of f_1 and f_2—and preceding the frequency changer with a stage of tuned amplification centred on f_1.

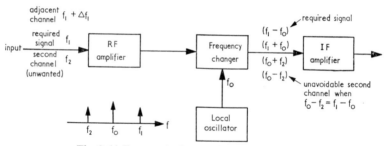

Fig. 3–11 Front end of a superheterodyne receiver.

This last statement requires further qualification. For the reception of signals in the LF, MF and lower HF ranges an intermediate frequency of 500 kHz is suitable. This would result in $f_1 - f_2 = 1$ MHz and therefore a single-stage tuned amplifier could be designed to provide considerably more gain at f_1 than at f_2. On the other hand, for VHF and UHF reception a 1 MHz separation could not be considered large, since both wanted and second channel inputs would experience the same order of amplification.

Next, let us consider the requirements for removing an *adjacent channel* $f_1 + \Delta f$. The inputs to the IF amplifier are $f_1 - f_0$ due to the wanted signal, and $f_1 + \Delta f - f_0$ corresponding to the adjacent channel. The lower the IF the higher the selectivity of the amplifier and the greater the rejection of the adjacent channel, but unfortunately this is contrary to the requirements for second-channel suppression. A solution is to use two stages of IF amplification, the first at a relatively high frequency for the purpose of second-channel rejection and the second at a low frequency for high selectivity. Typical values for HF reception are 1·6 MHz and 100 kHz for the first and second IFs respectively, whilst for VHF 10·7 MHz and 100 kHz are common and some UHF receivers use three stages at 50 MHz, 5–10 MHz and 100–500 kHz. It is to be stressed that these are only typical values and that many different ones are to be found in common use.

If the need to avoid second-channel and adjacent-channel interference were the only problems confronting the design engineer, his task would be a relatively simple one, but, as one may imagine, this is not the case. The frequency changer, and also non-linearities in the tuned amplifier preceding it, gives rise to spurious outputs due to *crossmodulation* between signals having amplitudes that are orders of magnitude greater than the required input. These are similar to the intermodulation products generated in a modulator (see page 45), but in the case of the frequency changer the interference is considerably worse, since the amplitude range (or dynamic range) of the input signals is much greater than the dynamic range of a baseband signal and therefore higher-order terms in the polynomial expansion of Eq. (3–1), i.e. dv^4, etc., make significant contributions.

It has been shown that spurious modulation is not generated, or at least it is extremely small, in a modulator that employs diodes operated in a piecewise-linear manner. There is, however, a drawback to using this form of modulator as a frequency changer. The *conversion gain*, i.e. the ratio of input signal power to output power at the required frequency, is very much less than unity, and it is shown in Chapter 4 that the gain must be as high as possible in the first and second stages of a receiver or amplifier. Consequently, the bipolar transistor operating as a small-signal common emitter amplifier is used in preference to diodes. The base current conforms to the polynomial expansion of Eq. (3–1) as for the diode, but the output, which is proportional to collector current, experiences the current gain of the device. The disadvantage is the generation of cross-modulation components. For this reason the bipolar transistor as a frequency changer has been superseded by the unipolar transistor, such as the FET and MOST[8, 9] whose input voltage/output current characteristics conform closely to the ideal square-law response.

8. Non-Linear Distortion and Mixing Processes in Field-Effect Transistors, J. S. Vogel, *Proc. I.E.E.E.*, Vol. 55, No. 12, pp. 2109–16, December 1967.
9. Application of Dual-gate MOS Field-Effect Transistors in Practical Radio Receivers, H. M. Kleinman, *Trans. I.E.E.E.*, Vol. BTR–13, No. 2, pp. 72–81, July 1967.

There are, of course, spurious contributions due to higher-order terms, but these are several orders of magnitude less than those of a bipolar device.

Having outlined the basic principles of signal amplification and filtering, we are now in a position to consider the demodulation process.

3–2.1 *Large-Signal Demodulation of a Full AM Signal ('Linear' Envelope Detection)*

It has been suggested on page 54 that the baseband information can be recovered by rectifying the full AM signal to produce an output that is proportional to its envelope. If the peak-to-peak amplitude of the signal from the IF amplifier is of the order of volts, rather than millivolts, a diode used as the rectifier can be considered as operating in a piecewise-linear mode, see Fig. 3–12(*a*)—refer also to the large-signal modulator of section 3–1.2. It can be seen that the output consists of the envelope of the AM signal supported by half-sinusoids at the carrier frequency. The input to the detector is

$$[E_c+m(t)]\cos\omega_c t$$

and the output can be expressed as

$$[E_c+m(t)] \times \frac{1}{\pi}\left[1+\frac{\pi}{2}\sin\omega_c t-\frac{2}{1\times3}\cos2\omega_c t-\frac{2}{3\times5}\cos4\omega_c t- \ldots\right]$$

where the Fourier series expansion represents the half-sinusoids. We see that the output contains the baseband information $m(t)$ together with higher-frequency terms involving f_c, $2f_c$, $4f_c$, etc. If the carrier frequency is very much higher than the highest frequency in the baseband, the unwanted components can be removed by a simple filter consisting of a capacitor in shunt with the load resistor.

3–2.2 *Small-Signal Demodulation of a Full AM Signal (Square-Law Detection)*

A second method of recovering the baseband information makes use of the non-linear characteristic of the diode described by Eq. (3–1), page 45. The input must have a peak-to-peak amplitude range of hundreds of millivolts rather than volts. The diode current is given by

$$\begin{aligned}
i(t) =\ & a[E_c+m(t)]\cos\omega_c t \\
& +b\{E_c+m(t)]\cos\omega_c t\}^2 \\
& +c\{[E_c+m(t)]\cos\omega_c t\}^3 \\
=\ & a[E_c+m(t)]\cos\omega_c t \\
& +\frac{b}{2}(1+\cos2\omega_c t)[E_c^2+2m(t)E_c+m^2(t)] \\
& + \text{terms due to } cv^3(t)
\end{aligned}$$

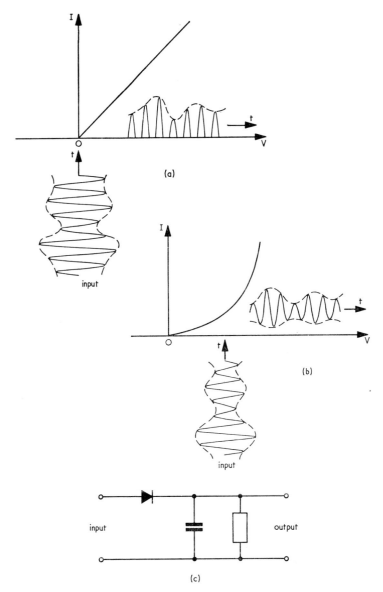

Fig. 3–12 Non-coherent demodulator. (a) Large-signal piecewise-linear diode characteristic, (b) square-law diode characteristic, and (c) circuit arrangement.

The reader should verify that the cv^3 term does not involve terms at baseband frequencies and therefore need not be considered any further. The required output is due to the bv^2 term—hence the description square-law detector—and it can be seen from the above expression that the baseband signal is recovered as $bE_cm(t)$. Unfortunately this is not the only contribution within the baseband spectrum, since the term $bm^2(t)/2$ will give rise to second harmonic distortion and intermodulation. To illustrate this point consider $m(t)$ to be made up of two sine waves $E_1\cos\omega_1 t$ and $E_2\cos\omega_2 t$, then

$$bm^2(t)/2 = b(E_1^2\cos^2\omega_1 t + 2E_1E_2\cos\omega_1 t\cos\omega_2 t + E_2^2\cos^2\omega_2 t)/2$$

and we see that there will be outputs at the second harmonic frequencies $2f_1$ and $2f_2$ and intermodulation components at (f_1+f_2) and (f_1-f_2). It is possible for all of these components to fall within the required baseband.

3–2.3 *Coherent Detection of DSBSC Signals*

A locally generated sine wave at the carrier frequency must be available before demodulation of a DSBSC signal can be accomplished. Not only

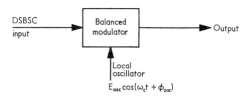

Fig. 3–13 Coherent demodulator for DSBSC signals.

is it essential for the sine wave to have the correct frequency, but its phase relative to the incoming signal is also important. Consider the DSBSC input* $m(t)\cos(\omega_c t+\phi_c)$ and the locally generated component $E_{osc}\cos(\omega_c t+\phi_{osc})$ to be applied to a multiplier (see Fig. 3–13). The output is

$$v_{out}(t) = m(t)\cos(\omega_c t+\phi_c)E_{osc}\cos(\omega_c t+\phi_{osc})$$
$$= \tfrac{1}{2}E_{osc}m(t)\cos(\phi_c-\phi_{osc}) \qquad (3\text{–}11)$$

which is the undistorted baseband signal, but having an amplitude dependent upon the cosine of the phase difference. When $\phi_c=\phi_{osc}$ the output is a maximum, and when a quadrature phase relationship exists

* Note that the demodulator is preceded by an IF amplifier and therefore ω_c refers to an intermediate frequency and not the incoming carrier frequency of the receiver input signal. Likewise, the locally generated sine wave is at the intermediate carrier frequency.

the output is zero regardless of the magnitude of $m(t)$ or E_{osc}. We note therefore that satisfactory demodulation of a DSBSC signal relies upon frequency coherence *and* phase coherence. The full significance of this statement will be appreciated after reading Chapter 5, which deals with the detection of signals in the presence of noise.

In practice the multiplication process can be achieved by any of the methods used for modulation (see sections 3–1.1 and 3–1.2). As an example, consider the chopper demodulator (or modulator) of the form shown in Fig. 3–2 (*b*), page 46. Let $E_{osc}\cos\omega_c t$ be the switching signal. To simplify the Fourier series representation of the switching action the phase term ϕ_{osc} is not included; the validity of the analysis is in no way affected, since the required output, as given by Eq. (3–11), depends upon the phase difference $\phi_c - \phi_{osc}$. The output from the chopper demodulator is given by

$$v(t) \propto m(t)\cos(\omega_c t + \phi_c)\left[\frac{1}{2} + \frac{2}{\pi}(\cos\omega_c t + \tfrac{1}{3}\cos 3\omega_c t + \ldots)\right]$$

$$= m(t)\cos\phi_c$$

$$+ \text{ terms involving } f_c, 3f_c, \text{ etc., which may be filtered out.}$$

Note that when phase synchronisation exists between the incoming signal and the switching sine wave $E_{osc}\cos\omega_c t$, $\phi_c = 0$ and the output is a maximum.

3–2.4 *Coherent Detection of SSB Signals*

It has been shown in Chapter 2 that the SSB form of modulation is the baseband signal translated in frequency by an amount f_c, and therefore the demodulation process need only reverse this spectral transposition; that is to say, the demodulator can be considered as a frequency changer. It is left to the student to check this statement by considering an SSB signal due to a baseband signal made up of a number of discrete sinusoids. In common with the requirement for demodulating DSBSC signals any of the multipliers described previously would make a suitable detector. However, it is to be stressed that the locally generated sine wave at the receiver must be at the correct frequency, a condition which in practice can only be realized by using crystal-controlled or frequency-synthesised oscillators. In general, phase coherence is also necessary, otherwise phase distortion would result; for speech transmission this is unimportant, since the ear is insensitive to this type of distortion, (see also page 105).

LC tuned oscillators do not have the necessary frequency stability

[3–2]

and therefore a receiver using this type of generator could only operate satisfactorily if information of the carrier frequency is contained in the transmission. This means that the DSBSC form could not be used and the signal must be either full AM, DSB with a pilot carrier—a carrier component at reduced magnitude but sufficient to enable an automatic frequency control (AFC) system at the receiver to keep the local oscillator at the correct frequency—or SSB with pilot carrier. The use of *LC* tuned receivers for the reception of SSB signals is of academic interest only, since all fixed point-to-point services and the vast majority of mobile services employ crystal-controlled oscillators. On the other hand, broadcast services for domestic entertainment must use modulation systems that only require relatively inexpensive receivers for their reception and demodulation; consequently, full AM and FM are the preferred systems.

3–3 Frequency Modulation

A distinction can be made between the frequency modulation of a carrier derived from

    (i) a variable frequency oscillator, and
    (ii) a crystal controlled source.

For signals of the first classification, and at carrier frequencies of less than 500 MHz, say, *LC* tuned circuits are invariably used. The modulating signal is made to control the inductance or effective capacitance of the resonant circuit and thereby determine the instantaneous frequency. The generation of FM at higher frequencies can be carried out directly at the carrier frequency by using thermionic devices such as the Klystron.[10] Another method is to generate the signal at a lower frequency, i.e. less than 500 MHz, and then adopt a process of frequency translation.

Frequency modulation of a carrier derived from a crystal oscillator can only be achieved by an indirect means, since it is not possible to control the frequency of the resonant circuit by direct application of the modulating signal. Nowadays, nearly all communication systems use crystal-controlled sources for generating the carrier. The formation of a FM signal from a variable frequency oscillator finds application in the field of instrumentation and therefore an account of the basic principle of varying the inductance or capacitance of an oscillator resonant circuit is given below.

10. The Design of a Reflex-Klystron Oscillator for Frequency Modulation at Centimetre Wavelengths, A. F. Pearce and B. J. Mayo, *Proc. I.E.E.*, Vol. 99, Part IIIA, pp. 445–54, also A. H. Beck and A. B. Cutting, *ibid*, pp. 357–66, 1952.

3–3.1 *Generation of FM by varying the Inductance or Capacitance of an Oscillator Resonant Circuit*

The unmodulated resonant frequency f_0 of a shunt LC circuit is

$$f_0 = \frac{1}{2\pi\sqrt{LC}}$$

If the capacitance is considered to be a variable with a maximum change of $\pm \Delta C$, then

$$f_{min} = \frac{1}{2\pi\sqrt{L(C+\Delta C)}} \qquad (3\text{–}12)$$

and

$$f_{max} = \frac{1}{2\pi\sqrt{L(C-\Delta C)}} \qquad (3\text{–}13)$$

where f_0-f_{min} and $f_{max}-f_0$ correspond to the peak frequency deviation. Rearranging Eq. (3–12) gives

$$f_{min} = \frac{1}{2\pi\sqrt{LC}}\left(1+\frac{\Delta C}{C}\right)^{-\frac{1}{2}}$$

and if $\Delta C \ll C$,

$$f_{min} \approx \frac{1}{2\pi\sqrt{LC}}\left(1-\frac{\Delta C}{2C}\right)$$

$$= f_0\left(1-\frac{\Delta C}{2C}\right)$$

or

$$f_0-f_{min} \approx f_0\frac{\Delta C}{2C} \qquad (3\text{–}14)$$

Similarly,

$$f_{max}-f_0 \approx f_0\frac{\Delta C}{2C} \qquad (3\text{–}15)$$

Therefore the peak frequency deviation, or in general the instantaneous frequency deviation, is directly proportional to the change in capacitance provided that this is small compared with the total capacitance in the resonant circuit. By a similar analysis, but with the inductance being treated as a variable, the peak frequency deviation is given by

$$f_{max}-f_0 \approx f_0\frac{\Delta L}{2L} \qquad (3\text{–}16)$$

To generate a frequency modulated signal as defined on page 32, i.e. a signal whose instantaneous frequency deviation is directly proportional to the instantaneous amplitude of the modulating signal, a variable reactance device exhibiting a linear reactance/applied voltage characteristic is required. Such a device does not exist, but a reasonable approximation can be obtained by using the variable capacitance properties of a varactor diode[11] operated over a restricted portion of its characteristic. The reactance valve[12] provides a method of simulating either a variable capacitor or inductor and has in the past been used with success. Ferrite-cored inductors have also been considered.[13]

3–3.2 *Indirect FM*

When the carrier is derived from a crystal-controlled source an indirect means of frequency modulation is necessary. The well-established method known as the Armstrong modulator makes use of the similarity between full AM and narrow-band angle modulation (see page 41). Consider the DSBSC output from a balanced modulator added to a

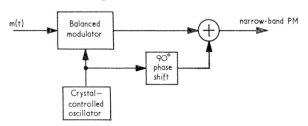

Fig. 3–14 Narrow-band phase modulation of a crystal-controlled carrier.

quadrature carrier component derived from the crystal oscillator (see Fig. 3–14). The output is effectively a narrow-band PM signal, since the phase displacement is directly proportional to the amplitude of the modulating signal $m(t)$. Note that this is only true for phase displacements not exceeding 0·2 rad. A narrow-band FM output is obtained by preceding the balanced modulator with an integrator. This will be appreciated by referring back to Eqs. (2–13) and (2–15), page 33.

Suppose we wish to form an FM signal conveying analogue information contained in a baseband of 50 Hz–15 kHz and to have a maximum frequency deviation of ±75 kHz and a carrier frequency within the VHF band, e.g. a BBC transmission in the range 87·5–100 MHz. The

11. Wideband FM with Capacitance Diodes, C. Arsem, *Electronics*, Vol. 32, No. 49, pp. 112–13, December 4, 1959.

12. *Frequency Modulation Engineering*, C. E. Tibbs and G. G. Johnstone, Chapter 7 (Chapman and Hall, 1956).

13. Variable Inductance Modulation of a Transistorized Sub-Carrier Oscillator, C. E. Land, *National Telemetering Conf., Denver, Colorado*, sponsored by A.I.E.E., pp. 78–102, May 1959.

peak phase displacement due to a component at the lowest frequency 50 Hz will be $\beta = 75$ kHz/50 Hz, which is very much greater than the allowable 0·2 rad of the narrow-band signal. It is therefore necessary to convert the latter, which incidentally will also exhibit a small amount of AM (see Fig. 2–12, page 42) into a wider-band version. This can be accomplished by using a non-linear circuit, fed with only the one input as distinct from a modulator or frequency changer requiring two, to generate a series of components harmonically related to the input. For example, if the input is a sine wave of frequency f, the output components from the *frequency multiplier* are at f, $2f$, $3f$, $4f$, etc. A suitable arrangement would be a diode operated as a switch, i.e. under large-signal conditions, and followed by a band-pass filter to extract the required harmonic. Another and better arrangement would be to use a transistor operating under similar large-signal conditions, since a higher conversion gain can be achieved and also the band-pass filter can be included as a tuned-collector load. These large-signal frequency multipliers also act as amplitude limiters and are therefore able to remove the unwanted AM.

As an illustration consider the narrow-band signal to have been formed at a lower carrier frequency, f_c' say, than the finally required output frequency. If this signal forms the input to a frequency multiplier, the output is

$$v_1(t) = E\cos n\left[\omega_c't + 2\pi \int_0^t k_f m(t)dt\right]$$

where n is the frequency multiplication factor and the input is expressed in the form given by Eq. (2–13), page 33. For ease of explanation, let us consider the modulating signal to be a single sine wave. The output from the multiplier becomes

$$v_1(t) = E\cos n[\omega_c't + \beta'\sin\omega_m t]$$

and $\beta' \leqslant 0·2$ rad. We see that the sub-carrier frequency f_c' and the peak phase displacement, or modulation index, are multiplied by n. Returning to the numerical example, the peak phase displacement of the required FM signal, corresponding to a 50 Hz component, is $\beta = 75 \times 10^3/50 = 1500$ rad/s. Since $\beta = n\beta'$, $n = 1500/0·2 = 7500$. If a sub-carrier frequency $f_c' = 100$ kHz is used, the resultant output carrier frequency would be 750 MHz, which is too high for our particular requirement. This may be overcome by interposing two frequency multipliers, producing the required multiplying factor n, i.e. $n_1 n_2 = n$, with a frequency changer that translates the signal to a lower frequency without any

[3–3]

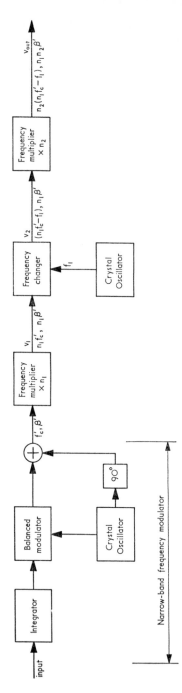

Fig. 3–15 Armstrong frequency modulator.

modification to the peak phase displacement. Consider the system shown in Fig. 3–15. The required output from the first frequency multiplier is of the form

$$v_1(t) = E\cos(n_1\omega_c't + n_1\beta'\sin\omega_m t) \qquad (3\text{–}17)$$

and the output from the following frequency changer is

$$v_2(t) = E\cos(n_1\omega_c't + n_1\beta'\sin\omega_m t)\cos\omega_1 t \qquad (3\text{–}18)$$

where ω_1 is the angular frequency of the crystal oscillator. The output at the difference frequency $f_c'' = (n_1f_c' - f_1)$ can be made the same order as f_c' by suitably choosing f_1. Note that in order to simplify the presentation the peak amplitude E is assumed constant throughout the process. The output from the frequency changer is an FM signal at a low subcarrier frequency f_c'', but with a modulation index $n_1\beta'$. The final output, from the second multiplier, has a frequency n_2f_c'' and a modulation index $\beta = n_1n_2\beta'$.

3–4 Demodulation of FM Signals

An FM detector (or discriminator as it is widely known) must produce an output voltage whose instantaneous amplitude is directly proportional to the instantaneous frequency of the input, i.e. it must possess the inverse characteristic of the modulator (see Eq. (2–9)). There are in general two ways of achieving this:

(i) by converting the frequency modulation into amplitude modulation, followed by envelope detection, or
(ii) by using pulse-counting techniques. With this method the FM signal is converted into a sequence of constant amplitude, constant duration pulses whose rate depends upon the instantaneous frequency of the input. Integration (or averaging) of the pulse train by a simple low-pass filter yields an output having an amplitude directly proportional to the pulse rate and hence to the instantaneous frequency of the FM signal.

Before demodulation can be carried out the wanted signal must be amplified and selected from other signals occupying the same transmission medium. The superheterodyne receiver, described earlier for the reception of AM, satisfies these requirements, but for FM reception the IF amplifier must also act as an amplitude limiter to remove any interfering AM which may have been caused by other signals mixing with the wanted input, noise, or distortion of the signal due to band-limiting effects of the earlier stages of the receiver. It will be readily

[3–4]

appreciated from the following sections that most discriminators are sensitive to amplitude changes of the input, and it is to be seen in Chapter 5 that the advantages of using FM in place of AM, when noise and interference are present, are fully gained when unwanted amplitude modulation is removed prior to demodulation.

3–4.1 *Discriminators employing an FM-to-AM Conversion*

When an FM signal is applied to a physically realizable network, the relative amplitudes and phases of the sidebands are modified, resulting in an output that is both amplitude and frequency modulated. If this signal is to be rectified by a diode acting as an envelope detector, the amplitude variations should, ideally, be identical to the original modulating signal. If they are not, the output from the detector will be distorted. In practice distortion is unavoidable and furthermore its theoretical assessment for a given network is not straightforward. Let us, however, define the network that will produce an undistorted AM signal from the FM input and hence lead to the faithful reproduction of the message.

Since the frequency of the signal is varying from one instant to the next, care must be exercised if the steady-state transfer function $H(j\omega)$ is to be used in determining the output response. To explain briefly, $H(j\omega)$ defines the input/output characteristic of a network in terms of its response to sinusoidal inputs of constant frequency. The transfer function can be recorded over any particular frequency range by varying the input frequency *slowly*, i.e. changing from one value to the next and establishing a steady-state condition at the new frequency before recording the output conditions of amplitude and phase. The response of a network to an input with rapidly changing frequency, as could be the case with FM, cannot be determined from the familiar steady-state relationship

$$v_{out} = H(j\omega)v_{in}$$

where v_{in} represents a varying frequency component, since the network may react sluggishly to a rapid change in frequency and this will not be evident from an inspection of $H(j\omega)$. It does not mean that the above expression cannot be used, but it does rule out the possibility of an exact solution by a substitution of Eq. (2–18), namely

$$e(t) = E_c\cos(\omega_c t + \beta\sin\omega_m t)$$

for the input v_{in}. However, an exact solution by steady-state analysis[14] can be obtained from the spectral representation of Eq. (2–21). Each

14. Non-Linear Distortion, with particular reference to the Theory of Frequency Modulated Waves, E. C. Cherry and R. S. Rivlin, *Phil Mag.*, Vol. 32, pp. 265–81, October 1941, and Vol. 33, pp. 272–93, April 1942.

sideband, which is at a *constant* frequency, is modified by the transfer function defining the network response at that particular frequency. The infinite series of Eq. (2–21) can be written as

$$e(t) = \sum_{n=-\infty}^{\infty} E_c J_n(\beta)\cos(\omega_c t + n\omega_m t) \qquad (3\text{–}19)$$

and the transfer function $H(j\omega)$ of the network as

$$H(j\omega) = A(\omega)\angle\phi(\omega)$$

with
$$A(\omega_c + n\omega_m) = A_n$$

and
$$\phi(\omega_c + n\omega_m) = \phi_n$$

where the subscript n signifies that the amplitude and phase factors define the response at the frequency corresponding to the nth sideband, i.e. at $\omega = \omega_c + n\omega_m$. The output is, therefore,

$$v_{out}(t) = \sum_{n=-\infty}^{\infty} E_c A_n J_n(\beta)\cos[\omega_c t + n\omega_m t + \phi_n] \qquad (3\text{–}20)$$

As an example, consider the network to have a linear amplitude-frequency characteristic and to offer zero phase shift. Then $A_n = a(\omega_c + n\omega_m)$, where a is the slope of the characteristic, see Fig. 3–16 (a). Eq. (3–20) becomes

$$v_{out}(t) = \sum_{n=-\infty}^{\infty} E_c a(\omega_c + n\omega_m) J_n(\beta)\cos[\omega_c t + n\omega_m t]$$

$$= a\omega_c E_c \sum_{n=-\infty}^{\infty} J_n(\beta)\cos[\omega_c t + n\omega_m t]$$

$$+ a\omega_m E_c \sum_{n=-\infty}^{\infty} n J_n(\beta)\cos[\omega_c t + n\omega_m t] \qquad (3\text{–}21)$$

The first term in the above equation is simply the original FM signal with a peak amplitude modified by the constant $a\omega_c$. In order to establish the significance of the second term, let us use the relationship.[15]

$$2n J_n(\beta) = \beta[J_{n+1}(\beta) + J_{n-1}(\beta)]$$

15. *Bessel Functions for Engineers*, N. W. McLachlan, p. 24 (Oxford University Press, 1946).

Therefore, $\displaystyle\sum_{n=-\infty}^{\infty} nJ_n(\beta)\cos[\omega_c t+n\omega_m t]$

$$= \sum_{n=-\infty}^{\infty} \frac{\beta}{2}\Big[J_{n+1}(\beta)+J_{n-1}(\beta)\Big]\cos[\omega_c t+n\omega_m t]$$

$$= \sum_{n=-\infty}^{\infty} \frac{\beta}{2}J_{n+1}(\beta)\cos[\omega_c t-\omega_m t+(n+1)\omega_m t]$$

$$+ \sum_{n=-\infty}^{\infty} \frac{\beta}{2}J_{n-1}(\beta)\cos[\omega_c t+\omega_m t+(n-1)\omega_m t]$$

combining the $(r+1)^{\text{th}}$ term of the first series with the $(r-1)^{\text{th}}$ term of the second series reduces the expression to

$$\sum_{n=-\infty}^{\infty} \beta J_n(\beta)\cos[\omega_c t+n\omega_m t]\cos\omega_m t$$

Therefore Eq. (3–21) becomes

$$v_{\text{out}}(t) = [\alpha(\omega_c+\beta\omega_m)\cos\omega_m t]E_c\cos(\omega_c t+\beta\sin\omega_m t) \qquad (3\text{–}22)$$

It can be seen that the frequency modulation of the output is the same as the input, but in addition there is amplitude modulation which gives rise to an envelope that is identical to the original modulating signal,

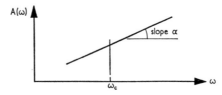

Fig. 3–16 Ideal linear amplitude-frequency characteristic of a network for FM-to-AM conversion.

since $\beta\omega_m=2\pi\Delta f$ and $\Delta f=k_f E_m$ (see page 34). Therefore, this simple but nevertheless unrealistic network satisfies the conditions postulated for zero distortion in the FM-to-AM conversion. These conditions are in no way violated by assuming a linear phase-frequency characteristic in place of zero phase shift. If we consider this alone, by making the amplitude-frequency characteristic a constant and equal to unity, say, the output is

$$v_{out}(t) = \sum_{n=-\infty}^{\infty} E_c J_n(\beta)\cos[\omega_c t + n\omega_m t + \psi(\omega_c + n\omega_m)]$$

where ψ is the slope of the characteristic (see Fig. 3–16). Rearranging gives

$$v_{out}(t) = \sum_{n=-\infty}^{\infty} E_c J_n(\beta)\cos[\omega_c(t+\psi) + n\omega_m(t+\psi)]$$

$$= E_c \cos[\omega_c(t+\psi) + \beta\sin\omega_m(t+\psi)]$$

and therefore the linear phase-frequency characteristic introduces no distortion but only a time delay.

When physically realizable networks having non-linear transfer functions are considered, the steady-state method of analysis can become extremely complicated, and often it is not possible to establish an exact solution. There are a number of alternative approaches; one is to evaluate an approximate solution by numerical analysis, another is the quasi-stationary method suggested by van der Pol,[16] or thirdly a dynamic analysis using the Convolution Integral.[17] It is felt that the last method should not be covered in an introductory textbook, as it is considered to be genuine post-graduate work; however, a brief description of the quasi-stationary approach is not out of place.

It has been shown by van der Pol that when the frequency deviation is small compared with the carrier frequency, the sluggishness of the network response to an input of changing frequency can be ignored and the output given by the approximation

$$v_{out}(t) \approx E_c A(\omega_i)\cos[\omega_c t + \beta\sin\omega_m t + \phi(\omega_i)] \qquad (3\text{–}23)$$

where ω_i defines the amplitude- and phase-frequency characteristics at the instantaneous frequency, i.e. $\omega_i = 2\pi f_c + 2\pi k_f E_m \sin\omega_m t$. In our present application we are only concerned with the amplitude variations of the output, since the envelope detection following the network is insensitive to frequency and phase changes of the carrier. The amplitude-frequency characteristic $A(\omega_i)$ can be written as

$$A(\omega_i) = A(\omega_c + [\omega_i - \omega_c])$$

16. The Fundamental Principles of Frequency Modulation, B. van der Pol, *J.I.E.E.*, Vol. 93, Part III, pp. 153–8, May 1946.

17. Theory of Low-Distortion Reproduction of FM Signals in Linear Systems, E. J. Baghdady, *Trans. I.R.E.*, Vol. CT-5, pp. 202–14, September 1958.

and because we have imposed the condition $\omega_i - \omega_c \ll \omega_c$, it is in order to express $A(\omega_i)$ as a Taylor series, i.e.

$$A(\omega_i) = A(\omega_c) + (\omega_i - \omega_c)\frac{dA(\omega_c)}{d\omega_i} + \frac{(\omega_i - \omega_c)^2}{2!}\frac{d^2A(\omega_c)}{d\omega_i^2} + \cdots$$

If we refer back to the linear amplitude-frequency characteristic of the network used in the example for the steady-state analysis, then

$$A(\omega_c) = a\omega_c \text{ and } \frac{dA(\omega_c)}{d\omega_i} = a$$

the slope of the characteristic; all higher derivatives are zero. Therefore,

$$v_{\text{out}}(t) \approx E_c[a\omega_c + 2\pi a k_f E_m\sin\omega_m t]\cos[\omega_c t + \beta\sin\omega_m t + \phi(\omega_i)] \quad (3\text{--}24)$$

and we see that the amplitude modulation is identical to that obtained by the steady-state analysis, see Eq. (3–22). In this example, of course, Eq. (3–24) is not an approximate solution but an exact one, since the idealised transfer function represents a network with no sluggishness and the condition $\Delta f \ll f_c$ need not apply.

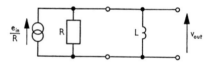

Fig. 3–17 Inductor fed from a constant-current source to produce an FM-to-AM conversion.

Let us now look at some physically realizable networks. The simplest arrangement, and, in fact, one that produces in theory zero distortion and in practice only a small amount, is an inductor fed from a constant current source (see Fig. 3–17). The output is

$$v_{\text{out}}(t) = \frac{e_{\text{in}}(t)}{R} j\omega L$$

assuming that $R \gg \omega L$, where R is the source impedance. The amplitude- and phase-frequency characteristics of the voltage transfer function are

$$A(\omega_i) = \frac{\omega_i L}{R} \text{ and } \phi(\omega_i) = \pi/2$$

and therefore, using Eq. (3–23), the output corresponding to an FM input is given by

$$v_{\text{out}}(t) \approx E_c(\omega_c + 2\pi k_f E_m\sin\omega_m t)\frac{L}{R}\cos(\omega_c t + \beta\sin\omega_m t + \pi/2) \quad (3\text{--}25)$$

The disadvantage of this simple network, and the reason for it not being used extensively in practice, is the extremely low efficiency of the FM-to-AM conversion. This can be seen from Eq. (3–25), bearing in mind that $\omega_i L \ll R$.

If ωL is allowed to be of the same order of magnitude as R,

$$A(\omega_i) = \frac{\omega_i L}{\sqrt{R^2 + \omega_i^2 L^2}}$$

and

$$v_{out}(t) = \frac{E_c[\omega_c + 2\pi k_f E_m \sin\omega_m t]L}{\{R^2 + (\omega_c + 2\pi k_f E_m \sin\omega_m t)^2 L^2\}^{\frac{1}{2}}} \cos[\omega_c t + \beta\sin\omega_m t + \phi(\omega_i)]$$

We see that the conversion efficiency is now much higher than that for the previous case, but unfortunately the frequency-dependent term in the denominator signifies distortion of the carrier envelope. A similar

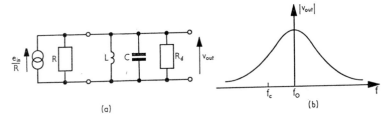

Fig. 3–18 (*a*) Off-resonance FM-to-AM converter. (*b*) Amplitude-frequency characteristic of the output.

phenomenon exists for the more widely used off-resonance circuit in which the resonant frequency f_0 is different from the carrier frequency f_c (see Fig. 3–18). The output

$$v_{out}(t) = e_{in}(t)\frac{Z(j\omega)}{R + Z(j\omega)} \qquad (3\text{–}26)$$

where $Z(j\omega) = R_d \Big/ \left[1 + jQ\left(\frac{\omega}{\omega_0} - \frac{\omega_0}{\omega}\right)\right]$

Using the result of the quasi-stationary approach, i.e. Eq. (3–23) and the transfer function of Eq. (3–26), it can be shown that the FM-to-AM conversion produces a carrier whose envelope variations are a distorted form of the modulating signal. If ω_c is far removed from ω_0 and the selectivity Q is high, the distortion can be made small. This, however, is not a practical solution, since the FM-to-AM conversion efficiency also

[3–4]

becomes small. It is not intended to carry out a quantitative analysis of distortion, as this should entail a dynamic analysis rather than the quasi-stationary method, and it has been stated previously that such an analysis is beyond the scope of this book.

The most widely used method of achieving a FM-to-AM conversion with high efficiency and relatively low distortion, compared with the

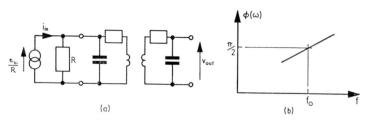

(a)   (b)

Fig. 3–19 (a) Mutually coupled parallel-tuned circuits. (b) Phase-frequency characteristic near resonance.

methods described above, makes use of the phase relationship between the input and output of two mutually coupled tuned circuits (see Fig. 3–19). It can be shown by general circuit theory that the input current $i(t)$ and the output voltage $v_{out}(t)$ are in quadrature[18] and that the phase-frequency characteristic $\phi(\omega)$ near resonance is approximately linear. The degree of linearity depends upon many factors including the Q, the

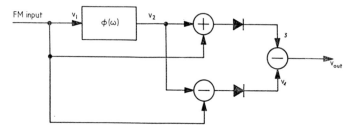

Fig. 3–20 Basic arrangement of a frequency discriminator employing a phase-dependent network.

coupling and, of course, the frequency range over which it is being considered. To simplify the analysis, we shall not only assume that the phase-frequency characteristic is linear but also that the amplitude-frequency characteristic is a constant. In practice any small departure from these conditions would result in distortion. Consider the schematic arrangement shown in Fig. 3–20.

18. See Reference 19.

The resonant frequency of the phase-sensitive network is made equal to the carrier frequency of the applied signal, i.e. $\phi(\omega_c) = \pi/2$. The input $v_1(t)$ corresponding to a frequency modulated signal is

$$v_1(t) = E_c\cos(\omega_c t + \beta\sin\omega_m t)$$

and the output from the phase-sensitive network is

$$v_2(t) = E_c\cos[\omega_c t + \phi(\omega)] \tag{3-27}$$

where
$$\phi(\omega) = \frac{\pi}{2} + k\Delta\omega\sin\omega_m t \tag{3-28}$$

and k is the slope of the phase-frequency characteristic. Note that this equation assumes the quasi-stationary analysis to be valid, i.e. the steady-state response and the dynamic response to be the same. Adding $v_1(t)$ and $v_2(t)$ gives

$$v_1(t) + v_2(t) = 2E_c\cos\left[\frac{\phi(\omega)}{2}\right]\cos\left[\omega_c t + \frac{\phi(\omega)}{2}\right]$$

and therefore the output from the envelope detector is

$$v_3(t) = 2E_c\cos[\phi(\omega)/2]$$
$$= E_c\sqrt{2}\sqrt{1+\cos\phi(\omega)} \tag{3-29}$$

Similarly $v_4(t) = E_c\sqrt{2}\sqrt{1-\cos\phi(\omega)}$ \qquad\qquad (3-30)

Subtracting Eq. (3-30) from Eq. (3-29) gives

$$v_{out}(t) = E_c\sqrt{2}\{\sqrt{1+\cos\phi(\omega)} - \sqrt{1-\cos\phi(\omega)}\}$$
$$= E_c\sqrt{2}\{\sqrt{1+\sin\theta} - \sqrt{1-\sin\theta}\}$$

where $\theta = \pi/2 - \phi(\omega)$. But $\phi(\omega)$ is in the region of 90°, therefore θ is small and the output is given by the approximation

$$v_{out}(t) \approx E_c\sqrt{2}\theta$$
$$= E_c\sqrt{2}[\pi/2 - \phi(\omega)]$$

and using Eq. (3-28).

$$v_{out}(t) \approx E_c\sqrt{2}k\Delta\omega\sin\omega_m t$$

and therefore the output is directly proportional to the frequency deviation. At carrier frequencies where lumped circuit techniques apply, the summations of $v_2 + v_1$ and $v_2 - v_1$ are usually achieved by parallel-

[3-4]

tuned transformers as in the Foster-Seeley and the Ratio-type discriminators.[19, 20] At higher frequencies, in the microwave region, the same basic principles have been applied using distributed circuits.[21]

3–4.2 *Pulse Counting Discriminator*[22, 23]

With this form of discriminator the FM signal is converted into a sequence of constant amplitude and constant duration pulses whose repetition rate depends upon the instantaneous frequency of the input. There are a number of ways of achieving this conversion: one such method employs a bistable Schmitt trigger to produce a square wave of varying p.r.f. To obtain pulses of constant duration the output from the Schmitt circuit is differentiated, and either of the resulting positive or negative going pulses are used to trigger a monostable circuit. Finally, a simple low-pass filter is used as an integrator to produce an output whose amplitude is directly proportional to the pulse rate and hence to the instantaneous frequency of the FM signal.

One of the main advantages of this arrangement is the wideband characteristic. It is possible to design a discriminator that will operate

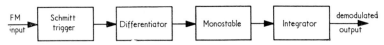

Fig. 3–21 Pulse-counter frequency discriminator.

satisfactorily for inputs within the frequency range 200 kHz–10 MHz. The second advantage is the extremely low level of distortion, due to a demodulation process that is essentially linear. The disadvantage, compared with discriminators employing an FM-to-AM conversion followed by envelope detection, is the higher output noise resulting from the additional components, some of which are active devices.

Further Tutorial Exercises

3–2 Single sine wave modulation of a carrier is used to form a full AM signal. One of the sidebands is suppressed by filtering and the remaining signal is fed

19. The Ratio Detector, S. M. Seeley and J. Avins, *RCA Review*, Vol. 8, No. 2, pp. 201–36, June 1947.

20. Foster-Seeley Discriminator, C. G. Mayo and J. W. Head, *Electronic and Radio Engr.*, Vol. 35, No. 2, pp. 44–51, February 1958.

21. *Frequency Modulation Theory*, J. Fagot and P. Magne, p. 310 (Pergamon Press, 1961).

22. Low-Distortion FM Discriminators, M. G. Scroggie, *Wireless World*, Vol. 62, No. 4, pp. 158–62, April 1956.

23. Limiters and Discriminators for FM Receivers, G. G. Johnstone, ibid., Vol. 63, No. 1, pp. 8–14, January 1957; No. 2, pp. 70–4, February 1957; No. 3, pp. 124–7, March 1957; No. 6, pp. 275–80, June 1957; No. 8, pp. 378–84, August 1957.

to an ideal envelope detector. If the original modulation index m_a is small, show that the second harmonic distortion of the output from the detector is approximately $12.5m_a\%$.

When the same signal is applied to a square-law detector show that the second harmonic distortion is zero.

3–3 The instantaneous voltage of a frequency modulated signal is given by
$$e = E_c\sin[10^7t + 37.5\sin(12.56 \times 10^3t)]$$
Determine the carrier frequency, the modulating signal frequency and the peak frequency deviation.

The volage e is applied across a series combination of a resistor R and capacitor C. If $\omega C \ll R$ in the frequency band occupied by the signal, show that the voltage developed across the resistor is amplitude modulated and calculate the depth of modulation.

Ans: 1·59 MHz; 2 kHz; 75 kHz; 4·7%.

3–4 The output from a 1 MHz LC tuned oscillator is frequency modulated by applying a modulating signal to the varactor diodes connected across the tuned circuit (see Fig. TE 3–1).

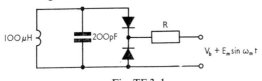

Fig. TE 3–1.

Frequency multiplication of the 1 MHz frequency modulated output results in an FM signal at a frequency of 100 MHz and with a modulation index $\beta = 10$.

Calculate the magnitude of the bias voltage V_b and the peak magnitude of the modulating sine wave $E_m\sin\omega_m t$, given that
   (i) $\omega_m = 2\pi \times 10^4$ rad/s;
   (ii) the LC circuit has L $= 100\mu$H and C $= 200$ pF, and
   (iii) the characteristic of each diode is
$$C = 200(V)^{-\frac{1}{2}}\text{pF}$$
   where V is the applied voltage.

Ans: 4V; 80 mV.

3–5 A periodic train of half sine waves can be expressed as
$$f(t) = \frac{1}{\pi}\left[1 + \frac{\pi}{2}\sin\frac{2\pi t}{T} - \frac{2}{1\times 3}\cos\frac{4\pi t}{T} - \frac{2}{3\times 5}\cos\frac{8\pi t}{T} - \cdots\cdots\right]$$

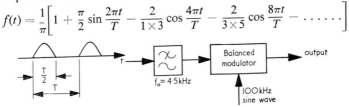

Fig. TE 3–2.

[3–4]

Calculate the bandwidth of the output signal from the system shown in Fig. TE 3–2 when the input signal has a period $T = 1$ millisecond.

> *Ans: 96 — 104 kHz.*

3–6 The instantaneous voltage of an amplitude modulated signal is given by

$$v = 0{\cdot}1(1+0{\cdot}2\sin300t+0{\cdot}35\sin700t)\sin10^6t \text{ volts}$$

Determine,

    (i) the r.m.s. amplitude of the signal;
    (ii) the peak depth of modulation, and
    (iii) the frequencies and relative amplitudes of the output components from an ideal square-law detector to which the signal has been applied.

> *Ans: 0·0735 V; 55%;*

rad/s	0	300	400	600	700	1000	1400
relative amplitude	1·081	0·4	0·07	0·02	0·7	0·07	0·06

4
NOISE

If noise did not exist, the transmission of information would be a *fait accompli*! But alas, this is not so and communication systems must be designed to keep the resulting equivocation of the received demodulated message to an acceptable level.

There are two main classes of noise, the first emanating from man-made sources and the second resulting from natural causes. Further subdivisions can be made: those of the first category are due either to wideband relatively low-power sources, such as car-ignition interference, radiation from fluorescent lights, domestic appliances, etc., or interference from adjacent channels occupying the same transmission medium; the last can, in general, be minimised by ensuring that each transmission only occupies its allocated bandwidth, having negligible spurious radiation over the rest of the frequency spectrum. The other man-made sources can be suppressed or, alternatively, avoided by housing the receiving terminal at a sparsely populated location.

All communication systems, whether line or radio, suffer from the effects of the random motion of electrons in conductors—a natural phenomenon referred to as *thermal noise* to indicate that the level is determined by the temperature of the conductor. Other natural sources are: outer space (*cosmic noise*), lightning (*electrostatic noise*) and the atmosphere acting as a black body due to the absorption of energy and its consequent re-radiation (*atmospheric absorption noise*).

To determine the effect of noise upon the intelligibility of a message we must endeavour to define it in mathematical terms. Since noise is random it is not possible to establish an algebraic expression which defines the amplitude-time dependence of a particular source, i.e. noise is a non-deterministic process. There are, however, two other methods of analysis both of which lend themselves to the study of system design.

The first method that we shall deal with involves the mean square amplitude of the noise, i.e. the noise power, instead of instantaneous values of amplitude. The reason for this approach is best illustrated by a simple experiment using a high-gain audio amplifier. With zero input signal, in fact with the input short-circuited, the output displayed on an

[4]

oscilloscope will be a randomly varying waveform. But the output can also be recorded as a steady deflection on a meter that responds to either the mean-square or r.m.s. amplitude; that is to say, for a given noise source the random fluctuations are averaged out to give a constant mean-square value. We can therefore talk about the output noise power of a system and consequently assess its effect upon the required signal in terms of a signal-to-noise power ratio—usually expressed in dB, see page 12. This method is particularly useful for the design and comparison of systems carrying analogue information. For example, the signal-to-noise power ratio at the receiver of a commercial telephone link should not fall below 26 dB, whereas for acceptable high-fidelity reproduction the ratio should be at least 50 dB.

The second method of defining a noise source makes use of its statistical properties. If these are known, it is possible to estimate the probability of the noise amplitude being a certain magnitude at any given time. This form of analysis lends itself to the study of digital communication systems in which it is necessary to estimate the average digit error rate (see Chapter 8). When using decision detection an error is recorded if the signal plus noise combine to produce a resultant amplitude that is the wrong side of the threshold level at the instant of making the decision.

Let us first consider thermal noise defined in terms of its mean-square amplitude.

4–1 Thermal Noise

The first thorough investigation of thermal noise was carried out in the 1920s by Johnson and Nyquist working independently. It was shown that the random motion of electrons in a conductor, due to thermal agitation, gives rise to a noise current such that all frequencies of the spectrum are represented in its variations—a characteristic resulting from the vast number of electrons involved. Because of its continuous spectrum extending over the whole of the frequency range of e-m waves, thermal noise is also known as *white noise* analogous to white light. The average noise current $\overline{i_n(t)}$ is zero, which is an intuitive statement when one considers that there is no applied e.m.f. and that the electron flow is equal in both directions. But Nyquist established, by applying the principles of statistical mechanics, that the mean-square current $\overline{i_n^2(t)}$ is not zero and is given by

$$\overline{i_n^2(t)} = 4kTGB \qquad (4\text{–}1)$$

where k is Boltzman's constant $(1 \cdot 37 \times 10^{-23} \text{J}/^\circ\text{K})$,

　　　　T is the absolute temperature of the conductor,

　　　　G is the conductance in siemens, and

B is that part of the frequency spectrum used for calculating the mean-square current—in practice it is the effective noise bandwidth of the device used for measuring $\overline{i_n^2(t)}$ (see page 91 for its definition).

We also note that $\overline{i_n^2(t)}$ is dependent upon the bandwidth *B*, but independent of the actual frequency, i.e. the position in the frequency spectrum. Consequently white noise has a uniform spectral power density.

The noise current in a conductor gives rise to an open-circuit voltage $v_n(t)$ across its terminals which has an average value of zero, but a mean-square value given by

$$\overline{v_n^2(t)} = \overline{i_n^2(t)}R^2$$

where $R = 1/G$ and is the resistance between the terminals. Using Eq. (4–1) gives

$$\overline{v_n^2(t)} = 4kTRB \qquad (4\text{--}2)$$

In order to apply circuit theory, the noisy resistor may be represented by an equivalent Thévenin voltage source or Norton current source, see Fig. 4–1(*a*), involving the appropriate noise generator and a

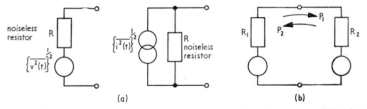

(a)   (b)

Fig. 4–1 (*a*) Equivalent voltage and current generators of a noisy resistor. (*b*) Exchange of power between two noisy resistors in parallel.

noiseless resistor. If either of these sources is connected to a resistive load, power will be absorbed, but, of course, if this load is a physically realizable resistor, it, too, must be represented as a noiseless resistor and a noise generator. When the two resistors are at the same temperature the net power flow is zero, but when they are at different temperatures, say T_1 and T_2 (see Fig. 4–1(*b*)), the resultant power flow is given by

$$P_1 - P_2 = \frac{4kR_1R_2B}{(R_1+R_2)^2}(T_1-T_2)$$

$$\left. \begin{array}{l} \text{where } P_1 = \dfrac{\overline{v_1^2(t)}}{(R_1+R_2)^2}R_2 \\[2ex] \text{and } P_2 = \dfrac{\overline{v_2^2(t)}}{(R_1+R_2)^2}R_1 \end{array} \right\} \qquad (4\text{--}3)$$

[4–1]

When $R_1 = R_2$ the power transfer from one side to the other is a maximum. Using Esq. (4–2) and (4–3) the maximum power (or *available power*, as it is more usually known) is given by

$$P_{1_{max}} = kT_1B \text{ and } P_{2_{max}} = kT_2B \qquad (4\text{–}4)$$

The above expression is based upon the derivation of Eq. (4–1) using the classical physics approach. In the light of quantum mechanics this equation is only approximately true, whereas the exact expression for available power is

$$kTB\left[\frac{hf}{kT}\left\{\exp\left(\frac{hf}{kT}\right)-1\right\}^{-1}\right]$$

where h is Planck's constant. When dealing with noise in systems working at frequencies of less than 30 GHz and ambient temperatures in the region of 290°K, the multiplying factor in squared brackets approaches closely to unity and can be ignored. Its inclusion may be warranted in cryogenic applications; as a useful rule of thumb the factor $hf/kT \approx f(\text{GHz})/20T(°K)$.

4–2 Shot Noise

The existence of thermal noise in a conductor does not rely on an applied e.m.f., i.e. there need not be an intentional current flow. In active

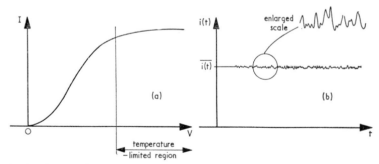

Fig. 4–2 (a) I-V characteristic of a thermionic diode. (b) Random variations about the average current $\overline{i(t)}$.

devices there is, apart from thermal noise, an additional contribution known as *shot noise*, due to the granular make-up of the current flow. It was first investigated using a thermionic diode operating under temperature-limited conditions, i.e. no space charge (see Fig. 4–2). Shottky[1] showed that the mean-square fluctuations are given by

$$\overline{i_n^2(t)} = 2e\,\overline{i(t)}B \qquad (4\text{–}5)$$

1. *Noise*, A. van der Ziel (Chapman & Hall, 1955).

where e is the electronic charge ($1 \cdot 6 \times 10^{-19}$ coulombs), and $\overline{i(t)}$ is the direct current (or average current).

Shot noise is in many ways similar to thermal noise. They are both due to the random fluctuations of a large number of electrons, have uniform spectral power densities, and furthermore the mean-square current in both cases is directly proportional to the bandwidth of the measuring instrument.

Although low-power thermionic diodes have been superseded by their semiconductor equivalents, the temperature-limited diode still fills a useful role as a calibrated noise source. A treatment of diodes working under space-charge-limited conditions is well documented elsewhere.[2]

Having outlined the two most common forms of naturally occurring noise, a reasonable continuation would be to consider various active devices and build models using thermal and shot-noise sources. However, instead of considering a specific device, such as the bipolar transistor, field-effect transistor, or travelling-wave tube, we shall consider them all as black boxes, paying little attention to the noise mechanisms, but investigating their effect upon the overall performance of a system—a justifiable approach, since we are laying the foundation to system design study.

4–3 Noise Figure of a Linear Network

The noise contribution of a linear network, whether it contains active or passive elements, can be assessed from a knowledge of its *noise figure F* which is defined as

$$F = \frac{signal\text{-}to\text{-}noise\ power\ ratio\ at\ the\ input}{signal\text{-}to\text{-}noise\ power\ ratio\ at\ the\ output}$$

$$= \frac{S_{\text{in}}}{N_{\text{in}}} \div \frac{S_{\text{out}}}{N_{\text{out}}} \qquad (4\text{--}6)$$

where S denotes signal power. It must be stressed that the noise-figure definition only applies to linear networks. A frequency changer (or mixer) can be considered as linear, since we have shown in Chapter 3 that it obeys the Superposition Theorem because of the one-to-one correspondence between input and output components. On no account can the characteristics of a non-coherent demodulator be defined in terms of a noise figure—this is dealt with in the next chapter.

2. *Information Transmission, Modulation and Noise*, M. Schwartz (McGraw-Hill, 1959).

[4–3]

A number of useful expressions follow from Eq. (4–6). If G is the power gain,

$$F = \frac{S_{in}N_{out}}{N_{in}GS_{in}} = \frac{N_{out}}{GN_{in}} \tag{4-7}$$

If the network were noiseless, $N_{out} = GN_{in}$ and its noise figure would be unity. Such a situation is physically unrealizable and in practice $N_{out} > GN_{in}$, i.e. $F > 1$.

Before continuing further let us consider, as an example, a single-stage amplifier. All the components will contribute to the additional noise appearing at the output, but it follows intuitively that noise developed at the front end of the amplifier, due to a resistor, say, will provide a much greater contribution to the total output noise than the same resistor in the output circuit of the amplifier. There is, in fact, a series of independent noise sources connected in tandem within the

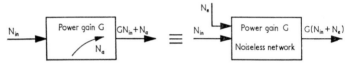

Fig. 4–3 Physically realizable network and its noiseless equivalent.

network that we are attempting to deal with as a black box. Nevertheless, for our present purposes we can lump all of these contributions together and refer to the total noise of the amplifier as N_a. Therefore Eq. (4–7) can be written as

$$F = \frac{GN_{in} + N_a}{GN_{in}} \tag{4-8}$$

It is desirable for ease of analysis to refer the amplifier noise N_a to the input terminals as an extra noise source N_e associated with a noiseless amplifier (see Fig. 4–3). Eq. (4–8) becomes

$$F = \frac{GN_{in} + GN_e}{GN_{in}}$$

$$= \frac{N_{in} + N_e}{N_{in}} \tag{4-9}$$

and rearranging gives

$$N_e = (F-1)N_{in} \tag{4-10}$$

We see that for a given network the effective input noise power N_e is not only exprsed in terms of the noise figure F but also expressed as a function of the noise from the source preceding it. The simplest arrangement is a resistor R_s connected across the input terminals (see Fig. 4-4(a)). Using Eq. (4-3), the input noise power $N_{in} = xkTB$ where $x = 4R_s R_{in}/(R_s+R_{in})^2$ and is unity when the input is matched to the source, i.e. $R_s = R_{in}$. Eq. (4-10) becomes

$$N_e = (F-1)xkTB \qquad (4-11)$$

and T $\approx$ 290°K when the resistor is at a normal ambient temperature. Note that the noise power flowing from the input of the network into the source has not been considered, since the noise power due to R_{in} is automatically included in the equivalent noise source N_e. For a given network, the left-hand side of Eq. (4-11) is a constant, and therefore the noise figure depends upon x, the matching conditions, as well as the temperature of the source.

Next consider the network to be preceded by an aerial. The latter can be represented as a generator of e.m.f. equal to the open-circuit noise

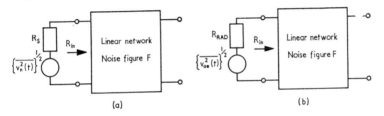

Fig. 4-4 Linear network preceded by (a) a noisy resistor, and (b) an aerial.

voltage appearing across its terminals, in series with its radiation resistance R_{RAD}, see Fig. 4-4(b). For the reader who is not familiar with basic aerial theory, it must be pointed out that R_{RAD} is not a physically realizable resistive component but an equivalent resistance accounting for the radiation properties of the aerial. It is probably most easily understood by considering the aerial as a radiator rather than as a receiver. Its radiation resistance is the resistance of an equivalent load capable of absorbing the same power as that radiated by the aerial.[3] The available noise power is $\overline{v_{ae}^2(t)}/4R$, where $R = R_{RAD} = R_{in}$ and if the noise is similar to thermal noise, which is certainly true of cosmic noise and a reasonable approximation for other types of e-m interference, then $\overline{v_{ae}^2(t)} \equiv 4kT_{ae}RB$ and hence the available noise power is $kT_{ae}B$; that is to say the available noise power from the aerial is equivalent to that from a resistor R_{RAD} at a temperature T_{ae}—this is known

3. *Electromagnetic Waves and Radiating Systems*, E. C. Jordan (Prentice-Hall, 1950, 10th edition, 1967).

as the *noise temperature* of the aerial. Typical values of T_{ae} for highly directional aerials operating in the SHF region pointing at a cold sky, i.e. not at a strong noise source, are to be found in Fig. 4–5.

Returning to the first arrangement of a source resistor R_s, it follows from the aerial work that, if R_s is the output resistance of an active device contributing shot noise and additional thermal noise, the available power to the network can be expressed as kT_sB where T_s is the noise temperature of the source.

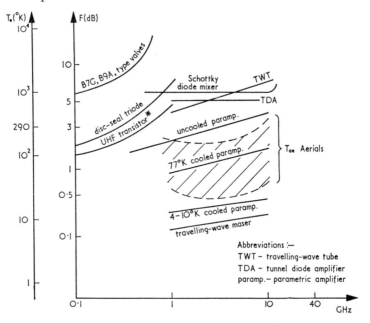

Fig. 4–5 Typical noise figure-frequency curves for various devices.†

The use of a noise temperature leads us to the alternative method of defining the noisiness of a network.

4–4 Effective Noise Temperature of a Linear Network

The need for a second method of defining network noise is best appreciated by enumerating the disadvantages of the noise-figure definition. First, we see from Eq. (4–9) that F is dependent upon the input noise N_{in}. In practice it is usually measured with the input of the network connected to a matched source resistor at 290°K, i.e. $N_{in} = 290kB$.

---

* Both noise figure and gain of a transistor should be specified at frequencies approaching cut-off; see *Microwave Journal*, Vol. 8, No. 1, p. 42, January 1965.

† See also, Syllabus on Low-noise Microwave Devices, N. E. Feldman, *Microwave Journal*, Vol. 12, No. 7, pp. 59–69, July 1969.

Strictly speaking this should always be stated when subsequently referring to the noise figure. It is to be seen from the work to follow that the equivalent noise temperature has the advantage of being independent of the source temperature. The second disadvantage is realised when attempting to compare modern low-noise devices in terms of noise figures that are all close to unity and within an extremely cramped scale. However, an expanded scale is provided by the equivalent noise temperature definition.

Referring back to the derivation of Eqs. (4–10) and (4–11), the extra noise power N_e may be considered to emanate from the same source resistance as N_{in}. Therefore we can write

$$N_e = xkT_iB \qquad (4\text{--}12)$$

where x describes the matching conditions and T_i is the increase in source temperature accounting for the noisiness of the network when the latter is replaced by its noiseless equivalent. But since N_e is a constant for a given network, T_i will depend upon x. Usually the source and network input resistances are matched, or considered to be matched, in which case N_e is the available power and is given by

$$N_e = kT_eB \qquad (4\text{--}13)$$

where T_e, the *equivalent noise temperature* of the network, is equal to T_i when $x = 1$. Substituting Eq. (4–13) in Eq. (4–11), again with $x = 1$, gives

$$T_e = (F-1)T_{in} \qquad (4\text{--}14)$$

Note that F in the above equation is the noise figure measured under matched input conditions with the noise source at a temperature T_{in}.

For example, when $T_{in} = 290°K$, $T_e = (F_{290}-1)290°K$.

4–5 Cascaded Networks

It is often desirable to establish an overall noise figure or an equivalent noise temperature of a system made up of a number of networks connected in cascade and defined individually. Suppose that three such networks have power gains G_1, G_2 and G_3 respectively and noise figures F_1, F_2, and F_3, all measured under similar input conditions of noise power N_{in}, e.g. a matched resistor at normal ambient temperature. The cascaded arrangement is preceded by a matched noise source N_{ae} which may be different from N_{in}.

[4–5]

We see from Fig. 4–6 that the output noise power is

$$G_3\{G_2[G_1N_{ae}+G_1(F_1-1)N_{in}]+G_2(F_2-1)N_{in}\}+G_3(F_3-1)N_{in}$$

and substituting this expression into Eq. (4–7) gives the overall noise figure

$$F = \frac{G_3\{G_2[G_1N_{ae}+G_1(F_1-1)N_{in}]+G_2(F_2-1)N_{in}\}+G_3(F_3-1)N_{in}}{N_{ae}G_1G_2G_3}$$

$$= 1+(F_1-1)\frac{N_{in}}{N_{ae}} + \frac{(F_2-1)}{G_1}\frac{N_{in}}{N_{ae}} + \frac{(F_3-1)}{G_1G_2}\frac{N_{in}}{N_{ae}} \qquad (4\text{–}15)$$

and for the special case of $N_{in} = N_{ae}$, i.e. the noise figures of the three networks measured under similar conditions to the input of the cascaded arrangement,

$$F = F_1 + \frac{F_2-1}{G_1} + \frac{F_3-1}{G_1G_2} \qquad (4\text{–}16)$$

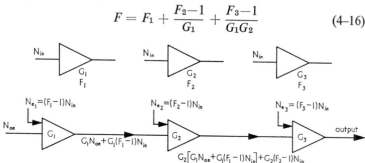

Fig. 4–6 Linear networks in cascade.

It can be seen from Eq. (4–16) that when the power gain of the first stage is high, 20 dB or more say, the overall noise figure is approximately equal to F_1. In practice the input stage of a receiving system should be chosen to have as high a gain as possible coupled with a low noise figure.

The overall noisiness of the cascaded system can also be expressed in terms of an equivalent noise temperature. Using Eq. (4–14), which in the present application is

$$T_e = (F-1)T_{ae}$$

where F is the overall noise figure given by Eq. (4–15), and T_{ae} is the noise temperature of the source, then

$$T_e = (F_1-1)\frac{N_{in}}{N_{ae}}T_{ae} + \frac{(F_2-1)}{G_1}\frac{N_{in}}{N_{ae}}T_{ae} + \frac{(F_3-1)}{G_1G_2}\frac{N_{in}}{N_{ae}}T_{ae}$$

But $N_{in} = kT_{in}B$, where T_{in} is the network source temperature at which the noise figure F_1 (or F_2 or F_3) is measured. Also $N_{ae} = kT_{ae}B$; therefore the overall noise temperature becomes

$$T_e = (F_1-1)T_{in} + \frac{(F_2-1)}{G_1}T_{in} + \frac{(F_3-1)}{G_1G_2}T_{in}$$

and using Eq. (4–14), we get

$$T_e = T_{e_1} + \frac{T_{e_2}}{G_1} + \frac{T_{e_3}}{G_1G_2} \tag{4–17}$$

where T_{e_1}, T_{e_2} and T_{e_3} are the equivalent noise temperatures of the three networks.

4–6 Relationship between the Noise Figure and Insertion Loss of a Passive Network

Although we have so far been dealing with a general network which can be either active or passive, there exists for the latter kind a relationship between the insertion loss and noise figure. Suppose we have a lossy network such as a piece of waveguide that is matched to the source and load resistances (see Fig. 4–7 (*a*)); the available noise power from the

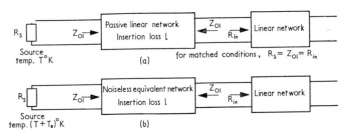

Fig. 4–7 (*a*) Passive linear network of insertion loss *l* matched to the source and load.
(*b*) The noiseless equivalent network with additional source temperature T_e.

network is *kTB*. The noisy network can be replaced by an additional noise source at the front end of its noiseless equivalent. Let its insertion loss *l* or gain *g* be defined as

$$l = 1/g \text{ with } 0 \leqslant g \leqslant 1 \text{ and } \infty \geqslant l \geqslant 1$$

From Fig. 4–7 (*b*) we see that the available noise power is $(gkTB+gkT_eB)$ which must be equal to *kTB*.

Therefore
$$gkB(T+T_e) = kTB$$

and
$$T_e = \left(\frac{1-g}{g}\right)T$$

or
$$T_e = (l-1)T$$
$$\tag{4–18}$$

Comparing this expression with Eq. (4–14),

$$F = l$$

The work of the last three sections is best illustrated by a numerical example.

Example 4–1. The ground receiver of a satellite-to-ground SHF communication link is shown in block diagram form in Fig. 4–8. The various stages may be considered matched for maximum power transfer. Calculate the equivalent noise temperature of the system at the waveguide input and thence calculate the signal-to-noise power ratio at the input to the demodulator, when the available signal power from the aerial is 7×10^{-13}W and the equivalent noise bandwidth of the receiver 10 MHz.

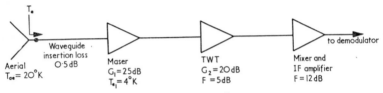

Fig. 4–8 SHF receiver.

Treat the system as a number of networks in cascade and use Eq. (4–17) to determine the equivalent noise temperature. First let us define each component in the system.

Waveguide: $l = 0.5$ dB

   therefore, $l = $ antilog $(0.05) = 1.12$

   and $G_1 = 1/l = 0.893$

Using Eq. (4–18), the equivalent noise temperature of the waveguide is

$$T_{e_1} = (1.12-1) \, 290 = 35°K$$

Maser: $T_{e_2} = 4°K$ given.

   and $G_2 = 25$ dB $\equiv$ power ratio of 316

   TWT: $G_3 = 20$ dB $\equiv$ power ratio of 100

   $F_3 = 5$ dB $\equiv$ power ratio of 3.16

Using Eq. (4–14) to find T_{e_3},

$$T_{e_3} = (3.16-1) \, 290 = 627°K$$

Mixer and IF amplifier:

   $F_4 = 12$ dB $\equiv$ power ratio of 15.8

therefore, $T_{e_4} = (15.8-1) \, 290 = 4290°K.$

Using Eq. (4–17), the overall equivalent noise temperature referred to the waveguide input is

$$T_e = 35 + \frac{4}{0.893} + \frac{627}{0.893 \times 316} + \frac{4290}{0.893 \times 316 \times 100}$$

$$= 35 + 4.48 + 2.22 + 0.152$$

$$= 41.85°K$$

The total effective noise temperature $= T_a + T_e$
$$= 20 + 41 \cdot 85$$
$$= 61 \cdot 85°\text{K}$$

The available noise power is $k(T_a + T_e)B$
$$= 1 \cdot 37 \times 10^{-23} \times 61 \cdot 85 \times 10^7$$
$$= 8 \cdot 54 \times 10^{-15}\text{W}$$

Therefore the signal-to-noise power ratio at the output of the system is $7 \times 10^{-13}/8 \cdot 54 \times 10^{-15} = 82 \equiv 19 \cdot 1$ dB.

4–7 Statistical Properties of White and Band-limited Noise

Apart from their application to the study of digital systems, as mentioned earlier in the introduction to this chapter, the statistical properties of noise can be used to establish a pseudo-deterministic function representing the amplitude-time variations of the output noise of a network. It is this aspect that is dealt with next, and its application for studying analogue demodulation methods is covered in Chapter 5.

White noise is typical of naturally occurring phenomena involving a vast number of independent and randomly varying contributions. The probability of a particular amplitude (either voltage or current) occurring at any instant in time is given by the *Normal* (or *Gaussian*) probability density function

$$p(x) = \frac{1}{\sigma\sqrt{2\pi}} \exp\left[-\frac{(x-\overline{x})^2}{2\sigma^2}\right] \qquad (4\text{–}19)$$

where σ is the standard deviation or r.m.s. amplitude fluctuations of the noise and $\overline{x}$ is the average value or d.c. level (see Appendix II, page 251).

Consider this noise to have a one-sided* uniform spectral power density η watts/Hz and to be applied to a physically realizable network whose transfer function is $H(j\omega)$. The one-sided spectral power density of the output $G(f)$ is given by

$$G(f) = \eta |H(j\omega)|^2 \text{ watts/Hz} \qquad (4\text{–}20)$$

and the total noise power N by

$$\left. \begin{array}{l} N = \eta \displaystyle\int_0^\infty |H(j\omega)|^2 df \\[4mm] N = \dfrac{\eta}{2} \displaystyle\int_{-\infty}^\infty |H(j\omega)|^2 df \end{array} \right\} \text{ watts} \qquad (4\text{–}21)$$

or

The output noise can also be described in terms of the *equivalent noise bandwidth* B_n of a hypothetical network having zero loss within the

---

* See Appendix I for the definition of one-sided and two-sided spectra.

[4–7]

passband and infinite loss outside it such that $N = \eta B_n$. Although the output noise is severely band-limited compared with the input, it exhibits a Gaussian distribution, since it is comprised of a very large number of randomly varying contributions.

Our aim is to construct a mathematical model that will enable the signal plus band-limited noise to be treated in algebraic form. Let the noise fluctuations within a finite bandwidth δf, where $\delta f \ll B_n$, be represented as a noise sinusoid

$$\sqrt{2G(f)\delta f}\cos[\omega t + \theta(t)]$$

The peak amplitude $\sqrt{2G(f)\delta f}$ accounts for the noise power in the bandwidth δf. This expression must be an approximation, since it is impossible to extract a single sine wave of noise from a continuous spectrum. The randomly varying phase angle $\theta(t)$ is included in the expression in an effort to make the noise model as realistic as possible by allowing the frequency of the sine wave to vary and hence represent all frequencies in the band δf; the rate at which $\theta(t)$ varies depends upon δf (cf. FM and PM). The total noise voltage $v_n(t)$ can be expressed as the summation of a large number of these discrete sinusoids, i.e.

$$v_n(t) = \sum_k \sqrt{2G(f)\delta f}\cos[\omega_k t + \theta_k(t)] \tag{4-22}$$

Since our aim is to represent a signal plus noise in algebraic form, it is convenient to write Eq. (4–22) as

$$v_n(t) = \sum_k \sqrt{2G(f)\delta f}\cos[(\omega_k - \omega_c)t + \theta_k(t) + \omega_c t]$$

$$= \sum_k \sqrt{2G(f)\delta f}\{\cos[(\omega_k - \omega_c)t + \theta_k(t)]\cos\omega_c t$$

$$-\sin[(\omega_k - \omega_c)t + \theta_k(t)]\sin\omega_c t\}$$

$$= x(t)\cos\omega_c t + y(t)\sin\omega_c t \tag{4-23}$$

where $x(t) = \sum_k \sqrt{2G(f)\delta f}\cos[(\omega_k - \omega_c)t + \theta_k(t)]$

and $\qquad y(t) = -\sum_k \sqrt{2G(f)\delta f}\sin[(\omega_k - \omega_c)t + \theta_k(t)]$

Eq. (4–23) is a pseudo-deterministic representation of Gaussian-distributed noise. The two carrier components are modulated by randomly varying functions $x(t)$ and $y(t)$ which are made up of a large number of sinusoids and therefore have Gaussian distributions. When the output is narrow-band noise, i.e. $B_n \ll f_c$, $x(t)$ and $y(t)$ will be slowly varying compared with the carrier frequency. Before proceeding further it is advisable to check that the total noise power is represented correctly by the above equation. We have

$$\overline{x^2(t)} = \sum_k 2G(f)\delta f \times \tfrac{1}{2}$$

$$= \sum_k G(f)\delta f = N, \text{ the total noise power.}$$

Similarly $\qquad \overline{y^2(t)} = N$, and

$$\overline{v_n^2(t)} = \overline{x^2(t)} \times \tfrac{1}{2} + \overline{y^2(t)} \times \tfrac{1}{2} = N$$

which is the correct result and completes the check.

An alternative expression for the amplitude-time fluctuations of narrow-band noise can be obtained by combining the two random functions $x(t)$ and $y(t)$ to form a single random variable. Commencing with Eq. (4–23) and using the familiar geometrical transformation from Cartesian to polar coordinates, the output noise becomes

$$v_n(t) = r(t)\cos[\omega_c t + \phi(t)] \qquad (4\text{–}24)$$

where $\qquad\qquad r^2(t) = x^2(t) + y^2(t) \qquad\qquad (4\text{–}25)$

and $\qquad\qquad \phi(t) = \tan^{-1}y(t)/x(t) \qquad\qquad (4\text{–}26)$

This means that the noise is being represented as a single sinusoid of varying phase angle $\phi(t)$ and of varying peak amplitude $r(t)$ which defines the envelope of the resulting waveform. Although Eqs. (4–23) and (4–24) will suffice for the analysis which follows in Chapter 5, it is not out of place at this stage to investigate the difference between $r(t)$ and either $x(t)$ or $y(t)$. We know from the Gaussian probability density function that $x(t)$ and $y(t)$ can have positive and negative values, but according to Eq. (4–24) $r(t)$ defines the peak amplitude of the noise sinusoid and therefore cannot have a negative value—this is also apparent from Eq. (4–25). Let us see how this difference can be reconciled.

The noise voltage $v_n(t)$ is made up of two independent Gaussian functions, namely $x(t)$ and $y(t)$; therefore the probability of a certain noise amplitude occurring is dictated by the joint probability density function

$$p(x)\,p(y) = \frac{1}{2\pi N} \exp\left[\frac{-(x^2+y^2)}{2N}\right]$$

since $\qquad\qquad \sigma_x^2 = \overline{x^2(t)} = \sigma_y^2 = \overline{y^2(t)} = N$

The probability of x lying between x and $(x+dx)$ and y lying between

y and $(y+dy)$ is $p(x) p(y) dx dy$. This joint probability function can be restated in terms of the variables $r(t)$ and $\phi(t)$. Using the transformation from Cartesian to polar coordinates, it follows that $dx \, dy = r \, dr \, d\phi$ and therefore

$$p(x) p(y) \, dx \, dy = \frac{1}{2\pi N} \exp\left[\frac{-(x^2+y^2)}{2N}\right] dx \, dy$$

$$= \frac{1}{2\pi N} \exp\left(-\frac{r^2}{2N}\right) r \, dr \, d\phi$$

$$= \frac{r}{N} \exp\left(-\frac{r^2}{2N}\right) dr \, \frac{d\phi}{2\pi} \qquad (4\text{--}27)$$

Note that this function does not depend upon the actual value of ϕ but only upon $d\phi$, which means that all angles are equally likely, i.e. ϕ has a uniform distribution. It also follows that r and ϕ are independent variables. Since $x(t)$ and $y(t)$ are also independent variables defining the same noise function $v_n(t)$ as $r(t)$ and $\phi(t)$, we have

$$p(x) p(y) \, dx \, dy = p(r) p(\phi) \, dr \, d\phi \qquad (4\text{--}28)$$

where the definition of the right-hand side of the equation follows automatically from that given previously for the left-hand side. But $\phi(t)$ has a uniform distribution, i.e. $p(\phi) \, d\phi = d\phi/2\pi$. Therefore from Eqs. (4–27) and (4–28) we have

$$p(r) \, dr = \frac{r}{N} \exp\left(-\frac{r^2}{2N}\right) dr$$

or $p(r)$, the probability density function defining the envelope variations of the noise sinusoid, is given by

$$p(r) = \frac{r}{N} \exp\left(-\frac{r^2}{2N}\right) \qquad (4\text{--}29)$$

This is widely known as the Rayleigh probability density function.

The probability distribution $P(r > r_0)$, is the proportion of the time for which the peak amplitude of the noise sinusoid exceeds r_0 and is given by

$$P(r > r_0) = \int_{r_0}^{\infty} \frac{r}{N} \exp\left(-\frac{r^2}{2N}\right) dr \qquad (4\text{--}30)$$

It is common practice to express the variable r in terms of its r.m.s. value, which requires a substitution $r/\sqrt{N} = s$. Hence

$$P(s > s_0) = \int_{s_0}^{\infty} s \exp\left(-\frac{s^2}{2}\right) ds$$

$$= \exp\left(-\frac{s_0^2}{2}\right) \qquad (4\text{--}31)$$

which is illustrated in Fig. 4–9.

To summarise, band-limited noise having a Gaussian distribution can be represented in pseudo-deterministic form as a noise sinusoid with random but uniformly distributed phase and varying peak amplitude conforming to a Rayleigh probability distribution. The reader should note that although there is no constraint on the bandwidth relative to the centre frequency of the passband, the main application of this technique is for the analysis of signals in the presence of narrow-band noise, and it is for this condition that the concept of a noise sinusoid with slowly varying envelope and phase is particularly useful.

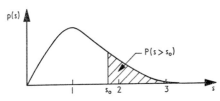

Fig. 4–9 Rayleigh probability distribution.

Tutorial Exercises

4–1 An amplifier is made up of three identical stages in tandem, each stage having equal input and output impedances. The power gain per stage is 8 dB, when correctly matched, and the noise figure is 6 dB.

Calculate the overall power gain and the noise figure of the amplifier.

Ans: 24 dB; 6·59 dB.

4–2 The noise figure of a receiver, without an aerial connected, is 0·9 dB measured at 290°K. Calculate the total effective noise temperature at the input of the receiver when an aerial of effective temperature 200°K is connected.

Ans: 266·6°K.

4–3 An aerial having an effective noise temperature of 50°K is connected, via a waveguide that has a loss of 0·8 dB, to a maser amplifier. The effective noise temperature of the amplifier is 4°K.

Calculate the total effective noise temperature at the input to the waveguide.

Ans: 112·8°K.

4–4 The input of a SHF receiving system consists of a high-gain aerial connected via a waveguide to a maser amplifier whose effective noise temperature is 4°K. The effective noise temperature of the system at the waveguide input increases from 55°K to 75°K due to ingress of moisture at a joint. Calculate the additional waveguide loss, in dB.

Ans: 0·24 dB.

4–5 A superheterodyne receiver is connected to an aerial having a noise temperature of 100°K by a length of coaxial line that has a total loss of 2 dB. The receiver characteristics are as follows:

Noise figure	4 dB
IF	20 MHz
IF bandwidth	1 MHz
RF bandwidth	5 MHz.

Calculate
   (i) the total system noise temperature, and
   (ii) the available received signal power from the aerial to give a 20 dB
        signal-to-noise ratio at the IF output.

Assume that the aerial and receiver are both matched to the coaxial line.
Boltzman's constant $= 1\cdot38 \times 10^{-23}$ J/°K.

Ans: 960°K; 1·332 $\times$ 10⁻¹² W.

4–6 Noise with Gaussian distribution, uniform spectral density and average
value $\overline{v(t)} = 0$ is applied to the input of the system shown in Fig. TE 4–1 (a).
The band-pass filter may be considered to have ideal characteristics and the
output filter to be its low-pass equivalent. The diode characteristic is shown
in Fig. TE 4–1 (b).

Determine the probability density function of the output and calculate its
r.m.s. amplitude if the probability distribution $P(v > 4 \text{ volts}) = 0\cdot13$.

Ans: 1·98 volts.

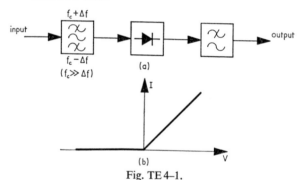

Fig. TE 4–1.

4–7 The circuit preceding an amplifier consists of a diode working under
temperature-limited conditions (see Fig. TE 4–2). When the direct current
through the diode is increased from zero to 11 mA the output noise power
from the amplifier is doubled. Calculate the noise figure of the amplifier.

Assume that the reactance of the blocking capacitor C is small and the
reactance of the RF choke is large compared with 50 ohms.

Ans: 3·1 dB.

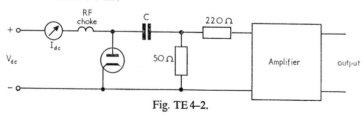

Fig. TE 4–2.

5
DEMODULATION OF ANALOGUE SIGNALS ACCOMPANIED BY BAND-LIMITED NOISE

The questions to be answered in this chapter are:

(i) How does the process of demodulation affect the signal-to-noise power ratio?

(ii) Is there any advantage in using coherent in place of non-coherent detection for the reception of AM?

(iii) How do AM and FM systems compare when carrying the same information and operating under similar conditions?

5–1 Non-coherent Detection of Full AM Signals

In the following sections it is shown how the input and output signal-to-noise power ratios for various AM demodulators are related, but before proceeding it is necessary to define the signal and noise bandwidths to which these ratios apply.

It was shown in Chapter 3 that a demodulator must be followed by a low-pass filter capable of removing all components other than those within the required baseband B_m. When there is negligible noise present the filter can be a simple RC network, since it is only necessary to separate the baseband signal components from those at carrier frequencies. If the input noise to the demodulator is significant, the low-pass filter should have a sharp cut-off characteristic to exclude all noise contributions outside the baseband B_m. Consider the input noise to be defined as $r(t)\cos[\omega_c t + \phi(t)]$ (see previous chapter). The output noise within the baseband B_m is due solely to the input sinusoid representing noise in the band $(f_c - B_m)$ to $(f_c + B_m)$—or, in other words, $r(t)$ is equivalent to a modulating noise signal of identical bandwidth to the message signal $m(t)$, producing a DSBSC sinusoid $r(t)\cos[\omega_c t + \phi(t)]$. Therefore the effective input noise power to an AM demodulator is the power associated with the output noise from a band-pass filter whose equivalent low-pass filter determines the demodulated output spectrum (see Fig.

[5–1]

5–1 and also Appendix I, page 238). Strictly speaking, the amplitude-frequency characteristics of these filters should be considered flat over their passbands, since in Chapter 4 we have assumed a uniform spectral density $G(f)$ for band-limited noise.

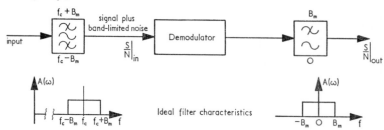

Fig. 5–1 General arrangement of filtering and demodulation of AM signals.

5–1.1 *Envelope Detection*

(see also Section 3–2.1, page 58)

The required baseband output is derived from the envelope of the signal when the latter is applied to a diode detector operating in a piecewise-linear mode. An expression for the input to the detector, consisting of signal plus band-limited noise, is obtained by combining Eqs. (2–3) and (4–24), giving

$$e_{in}(t) = [E_c + m(t)]\cos\omega_c t + r(t)\cos[\omega_c t + \phi(t)] \qquad (5-1)$$

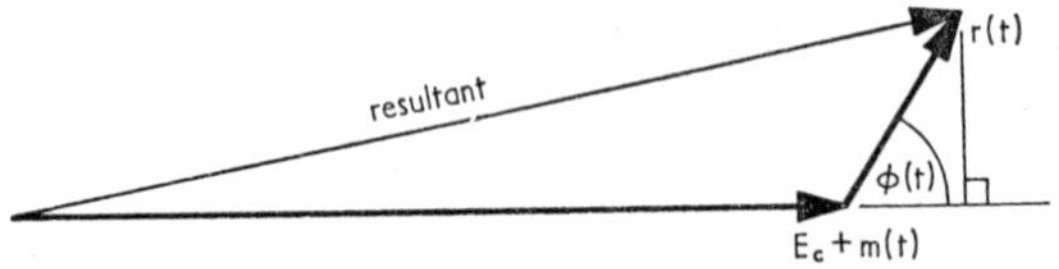

Fig. 5–2 Phasor diagram of full AM signal plus noise.

It can be seen from the phasor diagram of Fig. 5–2 that the envelope of the signal plus noise $e_R(t)$ is

$$e_R(t) = (\{[E_c + m(t)] + r(t)\cos\phi(t)\}^2 + \{r(t)\sin\phi(t)\}^2)^{\frac{1}{2}}$$
$$= \{[E_c + m(t)]^2 + 2r(t)[E_c + m(t)]\cos\phi(t) + r^2(t)\}^{\frac{1}{2}} \qquad (5-2)$$

It is convenient to consider the implications of Eq. (5–2) for two conditions of signal strength relative to the noise. Firstly, when

$[E_c + m(t)] \gg r(t)$, i.e. high signal-to-noise ratio, Eq. (5–2) can be written as

$$e_R(t) = [E_c + m(t)]\left\{1 + \frac{2r(t)\cos\phi(t(}{E_c + m(t)}\right\}^{\frac{1}{2}}$$
$$\approx [E_c + m(t)] + r(t)\cos\phi(t) \qquad (5-3)$$
$$\quad\quad\ signal \quad\quad\quad noise$$

It is worth noting that the same result is obtained when the signal plus noise is expressed as

$$e_{in}(t) = [E_c + m(t)]\cos\omega_c t + x(t)\cos\omega_c t + y(t)\sin\omega_c t$$

The quadrature noise component may be neglected when considering the envelope of the resultant input, and hence

$$e_R(t) \approx [E_c + m(t)] + x(t) \qquad (5\text{-}4)$$

which is identical to Eq. (5-3), since $r(t)\cos\phi(t) = x(t)$.

The second condition is for

$$r(t) \gg [E_c + m(t)]$$

i.e. low signal-to-noise ratio. Eq. (5-2) can be written as

$$e_R(t) = r(t)\left\{1 + \frac{2[E_c + m(t)]\cos\phi(t)}{r(t)}\right\}^{\frac{1}{2}}$$

$$\approx \underset{noise}{r(t)} + \underset{noise}{E_c\cos\phi(t)} + \underset{signal \times noise}{m(t)\cos\phi(t)} \qquad (5\text{-}5)$$

The conclusion to be drawn from Eqs. (5-3) and (5-5) is that the detector output is made up of signal and noise components which are *additive* when the input ratio is high, but are *multiplicative* when the ratio is low—a characteristic that is typical of all non-coherent detectors. The transition between the two states, or *threshold level* as it is known, is dealt with in section 5-5.

Our aim is to establish an expression relating the signal-to-noise ratio of the output to that of the input. It will be realised from an inspection of Eq. (5-5) that this is not possible for low input ratios, since the signal and noise components are inextricably mixed; all that can be said is that under these conditions there will be considerable degradation of the signal. On the other hand, when the signal is much larger than the noise, it is possible to interpret its effect upon the output signal in terms of a signal-to-noise ratio. Using Eq. (5-1) the input ratio is

$$\left.\frac{S}{N}\right|_{\substack{in \\ full\ AM}} = \frac{[E_c^2 + \overline{m^2(t)}]/2}{\overline{r^2(t)}/2} \qquad (5\text{-}6)$$

Using Eq. (5-3), the output signal-to-noise power ratio is

$$\left.\frac{S}{N}\right|_{\substack{out \\ env.\ det.}} = \frac{E_c^2 + \overline{m^2(t)}}{\overline{r^2(t)}/2} \qquad (5\text{-}7)$$

$$[5\text{-}1]$$

which together with Eq. (5–6) yields

$$\left.\frac{S}{N}\right|_{\substack{\text{out}\\\text{env. det.}}} = 2\left.\frac{S}{N}\right|_{\substack{\text{in}\\\text{full AM}}} \tag{5–8}$$

which means that demodulation has produced a 3 dB *improvement* in the signal-to-noise ratio. Regrettably this statement needs further qualification, since it is to be seen in Eq. (5–7) that the term $E_c{}^2$ in the numerator is a d.c. component corresponding to the rectified unmodulated carrier level and, therefore, should not be considered as an output signal. A more realistic output signal-to-noise ratio should contain only the terms due to the fluctuating message information, i.e.

$$\left.\frac{S}{N}\right|_{\substack{\text{out}\\\text{env. det.}}} = \frac{\overline{m^2(t)}}{\overline{r^2(t)}/2} \tag{5–9}$$

Combining with Eq. (5–6) gives

$$\left.\frac{S}{N}\right|_{\substack{\text{out}\\\text{env. det.}}} = \frac{\overline{m^2(t)}}{[E_c{}^2+\overline{m^2(t)}]/2}\left.\frac{S}{N}\right|_{\substack{\text{in}\\\text{full AM}}} \tag{5–10}$$

Unlike the previous result, Eq. (5–8), in which the d.c. component was included as an output signal, the difference between the input and output signal-to-noise power ratios can only be determined for a specific modulating signal.

Example 5–1. As a simple example, let $m(t) = E_m\sin\omega_m t$. Then $\overline{m^2(t)} = E_m{}^2/2$ and from Eqs. (5–6) and (5–9) we get

$$\left.\frac{S}{N}\right|_{\substack{\text{out}\\\text{env. det.}}} = \frac{m_a{}^2}{(1+m_a{}^2/2)}\left.\frac{S}{N}\right|_{\substack{\text{in}\\\text{full AM}}} \tag{5–11}$$

where m_a is the modulation index, see page 29. When $m_a = 1$ the output ratio is 1·75 dB less than the input ratio (see section 5–6).

Example 5–2. As a more realistic example consider the modulation to be due to a non-deterministic signal such as speech. It has been found that speech amplitudes have approximately an exponential distribution defined by the probability density function described on page 144. The probability of the amplitude exceeding four times the r.m.s. value is exceedingly small, $P[m(t) > 4\{\overline{m^2(t)}\}^{\frac{1}{2}}] \approx 3\times10^{-4}$, and we may therefore assume that the dynamic range for a given speaker is from $-4\{\overline{m^2(t)}\}^{\frac{1}{2}}$ to $+4\{\overline{m^2(t)}\}^{\frac{1}{2}}$, assuming that $\overline{m(t)} = 0$. The peak amplitude of the modulating signal must not exceed the peak carrier amplitude if over-modulation, and hence distortion, is to be avoided (see Chapter 2, page 29); this requires

$$E_c = 4\{\overline{m^2(t)}\}^{\frac{1}{2}}, \text{ i.e. } \overline{m^2(t)} = E_c{}^2/16$$

Substituting in Eq. (5–10) gives

$$\frac{S}{N}\bigg|_{\substack{\text{out} \\ \text{env. det.}}} \approx \frac{1}{8}\frac{S}{N}\bigg|_{\substack{\text{in} \\ \text{full AM}}}$$

which means that there is approximately a 9 dB reduction in the signal-to-noise ratio due to demodulation. In practice a degradation of this magnitude could only be tolerated when the signal-to-noise ratio is high, say 40 dB or more. This is, in fact, a reasonable requirement, since, by making $4\{\overline{m^2(t)}\}^{\frac{1}{2}} = E_c$, very nearly all of the amplitude range of the modulating signal is accommodated and therefore the system can handle speech of an exceptionally high quality, which, of course, demands a high signal-to-noise ratio.

When the signal-to-noise ratio is low, corresponding to the threshold level, quality is usually sacrificed at the expense of intelligibility by compressing the amplitude range of the modulating signal to allow a higher mean depth of modulation. One such system, developed successfully by the British Post Office and known as Lincompex,[1] employs a compandor to compress the amplitude range of the signal before modulation, and at the receiver, after demodulation, a complementary expandor to restore the amplitude range to its original form.

5–1.2 *Square-Law Detection* (see also Section 3–2.2, page 58)

When a diode detector is operated under small-signal conditions, the diode current $i(t)$ responsible for the demodulation to baseband is given by

$$i(t) = b\, v_{\text{in}}^2(t)$$
$$= b\{[E_c + m(t)]\cos\omega_c t + r(t)\cos[\omega_c t + \phi(t)]\}^2$$

when the input is a full AM signal plus band-limited noise. Expanding the above expression gives

$$i(t) = b\{\tfrac{1}{2}[E_c + m(t)]^2 + r(t)[E_c + m(t)]\cos\phi(t) + \tfrac{1}{2}r^2(t)\}$$
$$+ b\{[E_c + m(t)]r(t) + \tfrac{1}{2}r^2(t)\}\cos[2\omega_c t + \phi(t)]$$
$$+ b\{\tfrac{1}{2}[E_c + m(t)]^2\}\cos 2\omega_c t \qquad (5\text{–}12)$$

The baseband output voltage $v_{\text{out}}(t)$, is

$$v_{\text{out}}(t) = bR_L\{\tfrac{1}{2}[E_c + m(t)]^2 + r(t)[E_c + m(t)]\cos\phi(t) + \tfrac{1}{2}r^2(t)\} \quad (5\text{–}13)$$

In the following analysis the coefficient b is ignored, since both signal and noise are affected alike. Also the load resistance R_L is neglected and

1. The Lincompex System, R. L. J. Awcock, *Point-to-Point Telecommunications*, Vol. 12, No. 3, pp. 130–42, July 1968.

[5–1]

the mean powers are normalised by assuming the signal and noise voltages to be developed across a 1 ohm resistor. Expanding Eq. (5–13)

$$v_{out}(t) = \tfrac{1}{2}E_c{}^2 + m(t)E_c + \tfrac{1}{2}m^2(t) + r(t)[E_c + m(t)]\cos\phi(t) + \tfrac{1}{2}r^2(t) \quad (5\text{–}14)$$

The mean normalised *signal* power $= \{\tfrac{1}{2}E_c{}^2\}^2 + \overline{\{m(t)E_c\}^2} \quad (5\text{–}15)$

Note that the term $m^2(t)/2$ in Eq. (5–14) has not been included—it is this term that is responsible for second harmonic and intermodulation distortion and, strictly speaking, should be treated as a noise contribution; however, to simplify the analysis it will not be considered any further. From the fourth term on the right-hand side of Eq. (5–14), which is the result of multiplicative mixing of signal and noise,

the mean normalised *signal-noise* power $= \overline{\{r(t)[E_c + m(t)]\cos\phi(t)\}^2}$

$$= \overline{r^2(t)[E_c{}^2 + \overline{m^2(t)}]}/2 \quad (5\text{–}16)$$

since $\overline{\cos^2\phi(t)} = \tfrac{1}{2}$, all values of ϕ being equally likely. From the last term in Eq. (5–14),

the mean normalised *noise* power $= \overline{\{\tfrac{1}{2}r^2(t)\}^2}$

$$= \tfrac{1}{4}\overline{r^4(t)}$$

The random variable $r(t)$ has a Rayleigh distribution and it may be shown that[2] $\overline{r^4(t)} = 2\{\overline{r^2(t)}\}^2$; therefore

the mean normalised *noise* power $= \tfrac{1}{2}\{\overline{r^2(t)}\}^2 \quad (5\text{–}17)$

Not all of this noise power is due to a fluctuating output. The noise amplitude (voltage) term $r^2(t)/2$*, of Eq. (5–13), has an average value that is not equal to zero, i.e. $\overline{r^2(t)}/2 \neq 0$, indicating that there is a d.c. output due to noise alone. Consequently, Eq. (5–17) may be restated as

$$\tfrac{1}{2}\{\overline{r^2(t)}\}^2 = \tfrac{1}{4}\{\overline{r^2(t)}\}^2 + \tfrac{1}{4}\{\overline{r^2(t)}\}^2 \quad (5\text{–}18)$$

total noise   d.c. noise   fluctuating noise
power   power   power

The signal-to-noise power ratio involving only the fluctuating output components is

$$\frac{S}{N}\bigg|_{\substack{out\\ \text{sq. law det.}}} = \frac{\overline{\{m(t)E_c\}^2}}{\overline{r^2(t)[E_c{}^2 + \overline{m^2(t)}]}/2 + \{\overline{r^2(t)}\}^2/4} \quad (5\text{–}19)$$

2. *An Introduction to Statistical Communication Theory*, D. Middleton, p. 401 (McGraw-Hill, 1960).

* At first sight it may appear that this term has the dimensions of (volts)^2, but it must be remembered that the factor bR, which is being ignored, has the dimensions (volts)$^{-1}$.

and when $E_c+m(t) \gg r(t)$, i.e. high signal-to-noise ratio,

$$\left.\frac{S}{N}\right|_{\substack{\text{out}\\ \text{sq. law det.}}} \approx \frac{\overline{m^2(t)}E_c^2}{\overline{r^2(t)}[E_c^2+\overline{m^2(t)}]/2} \qquad (5\text{–}20)$$

When $E_c+m(t) \ll r(t)$, i.e. low signal-to-noise ratio

$$\left.\frac{S}{N}\right|_{\substack{\text{out}\\ \text{sq. law det.}}} \approx \frac{\overline{m^2(t)}E_c^2}{\{\overline{r^2(t)}\}^2/4} \qquad (5\text{–}21)$$

Inspection of Eqs. (5–20) and (5–21) shows that considerable degradation of the required output occurs when the input signal-to-noise power ratio is low.

As an example, consider a sine wave modulating signal $m(t) = E_m\sin\omega_m t$, with modulation index m_a. Using Eq. (5–6),

$$\left.\frac{S}{N}\right|_{\substack{\text{in}\\ \text{full AM}}} = \frac{E_c^2[1+m_a^2/2]}{\overline{r^2(t)}} \qquad (5\text{–}22)$$

and for *high* signal-to-noise power ratios, Eq. (5–20) gives

$$\left.\frac{S}{N}\right|_{\substack{\text{out}\\ \text{sq. law det.}}} \approx \frac{m_a^2 E_c^2}{\overline{r^2(t)}[1+m_a^2/2]} \qquad (5\text{–}23)$$

Combining Eqs. (5–22) and (5–23), we get

$$\left.\frac{S}{N}\right|_{\substack{\text{out}\\ \text{sq. law det.}}} \approx \frac{m_a^2}{[1+m_a^2/2]^2} \left.\frac{S}{N}\right|_{\substack{\text{in}\\ \text{full AM}}} \qquad (5\text{–}24)$$

for high signal-to-noise power ratios. As a numerical example, when $m_a = 0\cdot 8$ the power ratio is degraded by approximately 4·3 dB.

5–2 Coherent Detection of DSBSC Signals (see also Section 3–2.3, page 60)

It has been shown previously that a locally generated sine wave having the correct carrier frequency must be available for the demodulation of a DSBSC signal. When the input signal is accompanied by narrow-band noise, the output from the demodulator is

$$\{m(t)\cos(\omega_c t+\phi_c)+r(t)\cos[\omega_c t+\phi(t)]\}\,E_{osc}\cos(\omega_c t+\phi_{osc})$$

The required baseband signal is $\frac{1}{2}m(t)E_{osc}\cos(\phi_c-\phi_{osc})$

and the baseband output due to noise is $\frac{1}{2}r(t)E_{osc}\cos[\phi(t)-\phi_{osc}]$

$$[5\text{–}2]$$

The mean square noise output is $\frac{1}{4}\overline{r^2(t)}E_{osc}^2 \times \frac{1}{2}$

(note $\overline{\cos^2[\phi(t)-\phi_{osc}]} = \frac{1}{2}$, since all values of ϕ are equally likely).

The output signal-to-noise power ratio is given by

$$\frac{S}{N}\bigg|_{\substack{out \\ coh.\ det.}} = \frac{\overline{m^2(t)}E_{osc}^2\cos^2(\phi_c-\phi_{osc})/4}{\overline{r^2(t)}E_{osc}^2/8}$$

$$= \frac{2\overline{m^2(t)}\cos^2(\phi_c-\phi_{osc})}{\overline{r^2(t)}} \qquad (5\text{-}25)$$

The input signal-to-noise power ratio corresponding to a DSBSC signal is

$$\frac{S}{N}\bigg|_{\substack{in \\ DSBSC}} = \frac{\overline{m^2(t)}/2}{\overline{r^2(t)}/2} \qquad (5\text{-}26)$$

and therefore by substitution

$$\frac{S}{N}\bigg|_{\substack{out \\ coh.\ det.}} = 2\cos^2(\phi_c-\phi_{osc})\frac{S}{N}\bigg|_{\substack{in \\ DSBSC}} \qquad (5\text{-}27)$$

The requirement of phase coherence, i.e. $\phi_c = \phi_{osc}$, for maximum signal output has been discussed previously, but we now see that the process of demodulation has brought about a 3 dB improvement of the signal-to-noise ratio. Furthermore, this improvement is independent of the signal-to-noise ratio, which means that this type of demodulator does not exhibit a threshold effect.

5–3 Coherent Detection of SSB Signals (see also Section 3–2.4, page 61)

An analytic representation of a SSB signal can be obtained from the phase-shift method of generating the signal, namely

$$v(t)\bigg|_{SSB} = m(t)\cos\omega_c t - H[m(t)]\sin\omega_c t \qquad (5\text{-}28)$$

where $H[m(t)]$ is $m(t)$, but with a constant phase shift of 90° over the entire baseband—this is also widely known as the Hilbert transform of $m(t)$.[3, 4]

3. Hilbert Transforms and Modulation, F. F. Kuo and S. L. Freeny, *Proc. Nat. Electronics Conf.* (USA), Vol. 18, pp. 51–8, 1962.

4. Toward a Unified Theory of Modulation, H. B. Voelcker, *Proc. I.E.E.E.*, Vol. 54, No. 3, pp. 340–53, March 1966, and pp. 737–55, May 1966.

Consider the signal plus band-limited noise and a locally generated sine wave $E_{osc}\cos \omega_c t$ applied to a product detector. The demodulation process gives rise to

$$\{m(t)\cos\omega_c t - H[m(t)]\sin\omega_c t + r(t)\cos[\omega_c t + \phi(t)]\}E_{osc}\cos(\omega_c t + \theta)$$

where θ is the phase difference between the carrier that is used to generate the SSB signal and the locally injected carrier at the demodulator. The baseband signal from the above expression is

$$\tfrac{1}{2}m(t)E_{osc}\cos\theta - \tfrac{1}{2}H[m(t)]E_{osc}\sin\theta$$

The first term represents the desired output and the second term represents distortion due to the lack of phase synchronism. Ideally, when $\theta = 0$ the baseband signal is $m(t)E_{osc}/2$ and the phase distortion is zero. For television signals this would be an essential requirement, since the eye is sensitive to phase changes; the ear, however, is insensitive to phase distortion and hence phase coherence is not an essential requirement when demodulating SSB speech signals. As an illustration, when $m(t) = E_m\cos\omega_m t$, the baseband output is

$$\tfrac{1}{2}E_m E_{osc}\cos\omega_m t\,\cos\theta - \tfrac{1}{2}E_m E_{osc}\sin\omega_m t\,\sin\theta = \tfrac{1}{2}E_m E_{osc}\cos[\omega_m t + \theta]$$

We conclude that when $m(t)$ is made up of a series of sine waves representing a complex modulating signal, each component will suffer a constant phase shift. But the condition necessary to avoid phase distortion is a phase shift that is directly proportional to frequency.

Continuing the analysis on the assumption that either phase coherence can be achieved or that the modulation is due to speech, the noise output at baseband frequencies is $r(t)E_{osc}\cos[\phi(t) - \theta]/2$ and the baseband signal-to-noise power ratio is

$$\left.\frac{S}{N}\right|_{\substack{\text{out}\\ \text{coh. det.}}} = \frac{\overline{m^2(t)}E_{osc}^2/4}{\overline{r^2(t)}E_{osc}^2/8}$$

$$= \frac{\overline{m^2(t)}}{\overline{r^2(t)}/2} \tag{5-29}$$

But the input signal-to-noise power ratio is

$$\left.\frac{S}{N}\right|_{\substack{\text{in}\\ \text{SSB}}} = \frac{\overline{m^2(t)/2} + \overline{m^2(t)/2}}{\overline{r^2(t)}/2} \tag{5-30}$$

and therefore,

$$\left.\frac{S}{N}\right|_{\substack{\text{out}\\ \text{coh. det.}}} = \left.\frac{S}{N}\right|_{\substack{\text{in}\\ \text{SSB}}} \tag{5-31}$$

$$[5-3]$$

It is to be noted that apart from the signal-to-noise ratio being unchanged by demodulation there is also no threshold effect. It is not apparent from the above analysis that the pre-demodulation bandwidth need only be half that of a DSBSC system conveying the same information. Therefore, if the same signal power is available for both systems, we have

$$\frac{S}{N}\bigg|_{\substack{\text{in}\\\text{SSB}}} = 2\frac{S}{N}\bigg|_{\substack{\text{in}\\\text{DSBSC}}} \tag{5-32}$$

and consequently from Eqs. (5–27) and (5–31)

$$\frac{S}{N}\bigg|_{\substack{\text{out}\\\text{coh. det.}\\\text{SSB}}} = \frac{S}{N}\bigg|_{\substack{\text{out}\\\text{coh. det.}\\\text{DSBSC}}} \tag{5-33}$$

assuming that $\phi_c = \phi_{osc}$ for DSBSC demodulation. Although this shows that the two modulation systems are identical in terms of the output signal-to-noise power ratios, in practice SSB is preferred as firstly, only frequency coherence is required, whereas both frequency and phase coherence are necessary for demodulating a DSBSC signal and, secondly, only half the bandwidth is needed.

5–4 Demodulation of FM and PM Signals (see also Section 3–4, page 67)

Unlike the demodulation of full AM and DSBSC signals, the passband of the filter preceding the FM or PM discriminator is greater than twice the bandwidth of the output low-pass filter, as it must be sufficiently wide to accommodate all the significant sidebands of the signal. Consequently, the input noise power corresponds to the noise within this bandwidth, which, of course, will depend upon the modulation index and highest frequency of the modulating signal.

An FM or PM signal accompanied by band-limited noise can be represented as

$$e(t) = E_c\cos[\omega_c t + \phi_c(t)] + r(t)\cos[\omega_c t + \phi(t)]$$

where $\phi_c(t)$ describes the message modulation. The resultant signal will vary in both amplitude and phase, and since, in general, frequency-conscious and phase-conscious discriminators respond to amplitude changes of the input, it is desirable to remove these unwanted effects by preceding the discriminator with an amplitude limiter. The phase perturbations due to the noise are best illustrated by the phasor diagrams of Fig. 5–3; the two diagrams are essentially the same but Fig. 5–3 (a) is drawn for the condition $E_c \gg r(t)$, while the second conveniently represents the condition $r(t) \gg E_c$. From Fig. 5-3 (a) the instantaneous phase $\phi_R(t)$ of the resultant signal may be written as

$$\phi_R(t) = \omega_c t + \phi_c(t) + \tan^{-1} \frac{r(t)\sin[\phi(t) - \phi_c(t)]}{E_c + r(t)\cos[\phi(t) - \phi_c(t)]}$$

and when $E_c \gg r(t)$,

$$\phi_R(t) \approx \omega_c t + \underbrace{\phi_c(t)}_{message} + \underbrace{\frac{r(t)}{E_c}\sin[\phi(t) - \phi_c(t)]}_{noise} \qquad (5\text{-}34)$$

Using Fig. 5-3 (b), the instantaneous phase is

$$\phi_R(t) = \omega_c t + \phi(t) - \tan^{-1} \frac{E_c\sin[\phi(t) - \phi_c(t)]}{r(t) + E_c\cos[\phi(t) - \phi_c(t)]}$$

and when $r(t) \gg E_c$,

$$\phi_R(t) \approx \omega_c t + \underbrace{\phi(t)}_{noise} - \underbrace{\frac{E_c}{r(t)}\sin[\phi(t) - \phi_c(t)]}_{message \times noise} \qquad (5\text{-}35)$$

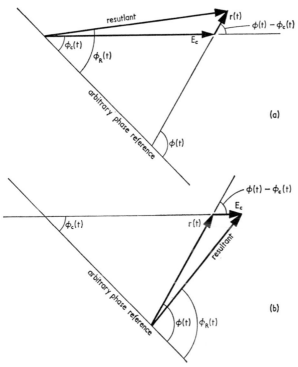

Fig. 5-3 Phasor diagrams of signal plus noise for (a) $E_c \gg r(t)$, and (b) $r(t) \gg E_c$.

[5-4]

Eqs. (5–34) and (5–35) illustrate the change from an additive signal and noise process for high signal-to-noise ratios to a multiplicative process when the ratio is low, a feature already discussed in section 5–1.1 when dealing with non-coherent AM demodulators.

The following analysis is only concerned with high signal-to-noise ratios, i.e. $E_c \gg r(t)$, for which the signal and noise outputs are additive, thus enabling the output signal-to-noise power ratio to be determined in two stages.

Stage 1. The noise is ignored and the output power of the required signal is determined.

If we consider the modulating signal to be a single sine wave, then

$$e(t)\Big|_{PM} = E_c\cos[\omega_c t + \beta\cos\omega_m t]$$

and

$$e(t)\Big|_{FM} = E_c\cos[\omega_c t + \frac{\Delta f}{f_m}\sin\omega_m t]$$

where β is the peak phase displacement and Δf the peak frequency deviation.

The mean output power from a phase discriminator	$\propto \frac{1}{2}\beta^2$	(5–36)
The mean output power from a frequency discriminator	$\propto \frac{1}{2}[\Delta f]^2$	(5–37)

For a more general treatment consider the PM signal to be

$$e(t)\Big|_{PM} = E_c\cos[\omega_c t + k_p m(t)]$$

and therefore

the mean output power from a phase discriminator	$\propto k_p^2\overline{m^2(t)}$	(5–38)

Similarly for FM, let

$$e(t)\Big|_{FM} = E_c\cos[\omega_c t + 2\pi\int_0^t k_f m(t)dt]$$

and hence

the mean output power from a frequency discriminator	$\propto k_f^2\overline{m^2(t)}$	(5–39)

Stage 2. This involves an approximation in which the message modulation is ignored and the output power is determined for an input consisting of an unmodulated carrier accompanied by a large number of randomly phased noise sinusoids. The approximation is not as unreasonable as it may appear at first sight, since the power of the unmodulated carrier is the same as that of the FM signal. This method does have a distinct advantage of simplicity; the significant output contributions are due solely to each noise sinusoid beating with the carrier—the effect of noise sinusoids mixing together to produce an output can be neglected, since the amplitudes of the carrier-noise components are much greater.

Consider each noise sinusoid to be of the form

$$\sqrt{2G(f)\delta f}\cos(\omega_n t + \theta_n)$$

See page 92 for an explanation of the terms. Each sinusoid combines with the carrier to produce a carrier-noise component having an instantaneous phase approximately equal to

$$\frac{\sqrt{2G(f)\delta f}}{E_c}\sin[(\omega_n - \omega_c)t + \theta_n]$$

Cf. Eq. (5–34). The instantaneous angular frequency perturbation is

$$\frac{d\,(phase)}{dt} \approx (\omega_n - \omega_c)\frac{\sqrt{2G(f)\delta f}}{E_c}\cos[(\omega_n - \omega_c)t + \theta_n]\ \text{rad/s}$$

The mean output power from a phase discriminator $\propto \frac{1}{2}\left[\dfrac{2G(f)\delta f}{E_c^2}\right]$

The mean output power from a frequency discriminator $\propto \frac{1}{2}(f_n - f_c)^2\left[\dfrac{2G(f)\delta f}{E_c^2}\right]$

The total output power is obtained by summing all the carrier-noise sinusoid contributions that produce outputs within the baseband B_m; that is to say, we are only concerned with the contributions for which $|\omega_n - \omega_c| \leqslant B_m$. Assuming that $\delta f \to df$, the total output noise power from a phase discriminator is proportional to

$$\frac{1}{E_c^2}\int_{f_c-B_m}^{f_c+B_m} G(f)df$$

It is reasonable to assume that $G(f)$ is constant over the band $(f_c - B_m)$ to $(f_c + B_m)$, see Fig. 5–4, and therefore

the total output noise power from a phase discriminator $\propto \dfrac{2B_m G(f)}{E_c^2}$ $\qquad$ (5–40)

[5–4]

Similarly,

the total output noise power from
a frequency discriminator

$$\propto \frac{1}{E_c{}^2} \int_{f_c-B_m}^{f_c+B_m} (f-f_c)^2 G(f) df$$

$$= \frac{2B_m^3 G(f)}{3E_c{}^2} \qquad (5-41)$$

It can be seen that in an FM system the output noise is not uniformly distributed throughout the baseband, since the noise power is directly proportional to $(f_n-f_c)^2$. As a result, the signal components at the

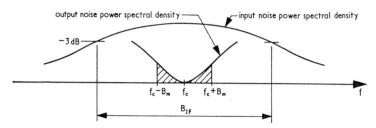

Fig. 5–4 Input and output noise power spectral densities associated with wideband FM discrimination.

higher frequencies suffer the most interference and in practice this is offset by emphasising the higher-frequency message components before modulating the carrier. A complementary de-emphasiser restores the signal to its correct form after demodulation.[5]

From Eqs. (5–38) and (5–40), the output signal-to-noise power ratio of a PM system is

$$\left.\frac{S}{N}\right|_{\substack{\text{out}\\ \text{PM}}} \approx \frac{k_p^2 \overline{m^2(t)} E_c{}^2}{2B_m G(f)} \qquad (5-42)$$

and multiplying the numerator and denominator by the equivalent noise bandwidth B_{IF} of the IF amplifier or pre-demodulation stage,

$$\left.\frac{S}{N}\right|_{\substack{\text{out}\\ \text{PM}}} \approx k_p^2 \overline{m^2(t)} \frac{B_{IF}}{B_m} \frac{E_c{}^2/2}{B_{IF} G(f)}$$

$$= k_p^2 \overline{m^2(t)} \frac{B_{IF}}{B_m} \left.\frac{S}{N}\right|_{\substack{\text{in}\\ \text{PM}}} \qquad (5-43)$$

5. *Frequency Modulation Engineering*, C. E. Tibbs and G. G. Johnstone, Chapter 4 (Chapman and Hall, 1956).

As a special case of single sine wave modulation,

$$k_p^2 \overline{m^2(t)} = k_p^2 E_m^2/2 = \beta^2/2$$

and therefore

$$\frac{S}{N}\bigg|_{\substack{\text{out} \\ \text{PM}}} \approx \beta^2 \frac{B_{IF}}{2B_m} \frac{S}{N}\bigg|_{\substack{\text{in} \\ \text{PM}}} \tag{5-43a}$$

Since Eqs. (5–43) and (5–43a) are the result of an approximate analysis, it is in order to assume that the equivalent noise bandwidth is equal to the 3 dB bandwidth of the IF amplifier.

From Eqs. (5–39) and (5–41), the output signal-to-noise power ratio of an FM system is

$$\frac{S}{N}\bigg|_{\substack{\text{out} \\ \text{FM}}} \approx 3\frac{k_f^2 \overline{m^2(t)}}{B_m^2} \frac{E_c^2}{2B_mG(f)} = 6\frac{k_f^2 \overline{m^2(t)}}{B_m^2} \frac{E_c^2/2}{2B_mG(f)} \tag{5-44}$$

(the equation is written in the form above for later use)

$$= 3\frac{k_f^2 \overline{m^2(t)}}{B_m^2} \frac{B_{IF}}{B_m} \frac{S}{N}\bigg|_{\substack{\text{in} \\ \text{FM}}} \tag{5-45}$$

and the special case of single sine wave modulation gives

$$k_f^2 \overline{m^2(t)} = k_f^2 E_m^2/2 = (\Delta f)^2/2$$

hence

$$\frac{S}{N}\bigg|_{\substack{\text{out} \\ \text{FM}}} \approx 3\left[\frac{\Delta f}{B_m}\right]^2 \frac{B_{IF}}{2B_m} \frac{S}{N}\bigg|_{\substack{\text{in} \\ \text{FM}}} \tag{5-45a}$$

A useful extension of the above analysis is to obtain the output signal-to-noise power ratio of a PM or FM system in terms of the corresponding output ratio of an AM system. Using Eq. (5–44), $2B_m$ is equal to the IF bandwidth required for an AM system and if equal signal power is available for all systems,

i.e. $E_{c_{FM}}^2 = E_{c_{AM}}^2 + \overline{m^2(t)}$, then

$$\frac{S}{N}\bigg|_{\substack{\text{out} \\ \text{FM}}} \approx 6\frac{k_f^2 \overline{m^2(t)}}{B_m^2} \frac{S}{N}\bigg|_{\substack{\text{in} \\ \text{AM}}} \tag{5-46}$$

and with Eq. (5–10), we get

$$\frac{S}{N}\bigg|_{\substack{\text{out} \\ \text{FM}}} \approx 6\left[\frac{k_f^2 \overline{m^2(t)}}{B_m^2}\right] \frac{[E_{c_{AM}}^2 + \overline{m^2(t)}]/2}{\overline{m^2(t)}} \frac{S}{N}\bigg|_{\substack{\text{out} \\ \text{AM} \\ \text{env. det.}}} \tag{5-47}$$

$$[5-4]$$

For single sine wave modulation,

$$\frac{S}{N}\bigg|_{\substack{\text{out} \\ \text{FM}}} \approx 3\left[\frac{\Delta f}{B_{\mathrm{m}}}\right]^2\left(\frac{2+m_{\mathrm{a}}{}^2}{2m_{\mathrm{a}}{}^2}\right)\frac{S}{N}\bigg|_{\substack{\text{out} \\ \text{AM} \\ \text{env. det.}}} \tag{5-47a}$$

Similarly for a comparison of PM with AM,

$$\frac{S}{N}\bigg|_{\substack{\text{out} \\ \text{PM}}} \approx 2k_{\mathrm{p}}^2\overline{m^2(t)}\,\frac{[E_{c_{\text{AM}}}^2+\overline{m^2(t)}]/2}{\overline{m^2(t)}}\,\frac{S}{N}\bigg|_{\substack{\text{out} \\ \text{AM} \\ \text{env. det.}}} \tag{5-48}$$

$$= \beta^2\left(\frac{2+m_{\mathrm{a}}{}^2}{2m_{\mathrm{a}}{}^2}\right)\frac{S}{N}\bigg|_{\substack{\text{out} \\ \text{AM} \\ \text{env. det.}}} \tag{5-48a}$$

for single sine wave modulation.

Although the above analysis has been developed for both FM and PM signals, in practice FM is invariably used, chiefly because of the simpler discriminators for its reception—PM discriminators require a locally generated phase reference which automatically puts them at a disadvantage; this is only true for analogue transmission and certainly does not aply for digital systems. See also Chapter 2, page 40.

Example 5-3. To illustrate Eqs. (5-47) and (5-47a) we shall consider the transmission of high-quality sound occupying a baseband $B_{\mathrm{m}} = 15$ kHz, by AM and FM. The peak frequency deviation of the FM signal is ± 75 kHz.

Firstly, for single sine wave modulation and a modulation index of unity, Eq. (5-47a) becomes

$$\frac{S}{N}\bigg|_{\substack{\text{out} \\ \text{FM}}} \approx 3\left[\frac{75}{15}\right]^2\frac{3}{2}\frac{S}{N}\bigg|_{\substack{\text{out} \\ \text{AM} \\ \text{env. det.}}}$$

indicating an *FM improvement* of approximately 20 dB over a full AM system using envelope detection.

Secondly, consider a non-deterministic modulating signal having the statistical properties that were described in Example 5-2, page 100. It was shown that for AM

$$\overline{m^2(t)} = E_{c_{\text{AM}}}^2/16$$

and, if similar conditions are imposed for the FM system by arranging the peak deviation of 75 kHz to correspond to a modulation amplitude of four times its r.m.s. value,

$$k_{\mathrm{f}}^2\overline{m^2(t)} = k_{\mathrm{f}}^2\left[\frac{E_{m_{\text{peak}}}}{4}\right]^2 = \frac{(\Delta f)^2}{16}$$

since $k_t E_{m_{peak}} = \Delta f$, the peak frequency deviation. Therefore, Eq. (5–47) becomes

$$\left.\frac{S}{N}\right|_{\substack{\text{out} \\ \text{FM}}} \approx \frac{6}{16}\left[\frac{75}{15}\right]^2\left[\frac{(1+1/16)/2}{1/16}\right]\left.\frac{S}{N}\right|_{\substack{\text{out} \\ \text{AM} \\ \text{env. det.}}}$$

$$\approx 75\left.\frac{S}{N}\right|_{\substack{\text{out} \\ \text{AM} \\ \text{env. det.}}}$$

indicating an FM improvement of approximately 19 dB. It must be remembered that this improvement can only be realised when the input signal-to-noise ratios are high, and this leads us to the next section dealing with a comparison of FM and AM at the threshold level.

5–5 Threshold Levels

5–5.1 *Threshold Level with Non-coherent AM Detection*

There is no distinct mathematical transition between the additive signal and noise process corresponding to high signal-to-noise power ratio and the multiplicative process when the ratio is low. But it has been found, by subjective analysis of the demodulated output, that the required signal is dominant when its *peak* amplitude exceeds four times the r.m.s. noise amplitude, both being measured at the input to the demodulator. Consequently the threshold level is defined by

$$\text{peak signal amplitude} \approx 4\sqrt{B_{IF}G(f)}$$

where $G(f)$ is the spectral power density of the band-limited noise and B_{IF} the noise bandwidth of the IF amplifier which precedes the demodulator. Therefore the mean signal-to-noise power ratio defining the threshold level is

$$\left.\frac{S}{N}\right|_{\substack{\text{in} \\ \text{AM} \\ \text{threshold}}} \approx \frac{(16/2)G(f)B_{IF}}{G(f)B_{IF}} = 8 \ (9 \text{ dB}) \qquad (5\text{–}49)$$

5–5.2 *Threshold Levels of FM Systems*

The demodulation of FM signals in the presence of noise gives rise to two threshold levels and, as for AM, both are found by subjective analysis. The first defines the input signal-to-noise power ratio for which the FM system becomes superior to an AM system using coherent detection and carrying the same information. This is found to occur

[5–5]

when the peak signal amplitude is approximately four times the r.m.s. noise amplitude as for AM. Therefore we have

$$\left.\frac{S}{N}\right|_{\substack{\text{in}\\\text{FM}\\\text{1st threshold}}} \approx 8 \ (9 \ \text{dB}) \qquad\qquad (5\text{--}50)$$

The second threshold is defined by the signal-to-noise power ratio that gives the full FM improvement over the corresponding AM system. This is found to be 2–3 dB greater than the signal-to-noise ratio defining the first threshold (see Fig. 5–5).

It is interesting to compare the signal-to-noise power ratio defining the first FM threshold with the input power ratio of the corresponding

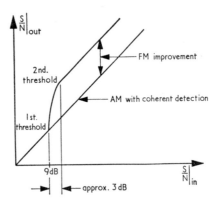

Fig. 5–5 Threshold levels of an FM system.

non-coherent AM system carrying the same information and having the *same* available signal power. Because the IF bandwidths of the two systems are different, we have the relationship

$$\left.\frac{S}{N}\right|_{\substack{\text{in}\\\text{AM}}} = \frac{B_{\text{IF(FM)}}}{B_{\text{IF(AM)}}} \left.\frac{S}{N}\right|_{\substack{\text{in}\\\text{FM}}}$$

The AM input ratio corresponding to the input ratio defining the first FM threshold is given by

$$\left.\frac{S}{N}\right|_{\substack{\text{in}\\\text{AM}}} = 8\,\frac{B_{\text{IF(FM)}}}{B_{\text{IF(AM)}}}$$

and since $B_{\text{IF(FM)}}$ is always greater than $B_{\text{IF(AM)}}$,

$$8\,\frac{B_{\text{IF(FM)}}}{B_{\text{IF(AM)}}} > \left.\frac{S}{N}\right|_{\substack{\text{in}\\\text{FM}\\\text{1st threshold}}}$$

indicating that at low signal-to-noise ratios AM gives a better performance than FM. This point is illustrated in the following section.

5–6 Comparison of Analogue Modulation Systems

A graphical representation of output signal-to-noise ratio plotted against input signal-to-noise ratio provides the most suitable form of comparison. It is assumed that all systems have equal signal power and the modulation is due to a single sine wave. Modulation by non-deterministic signals is not considered, but the reader is referred back to Examples 5–2 and 5–3 for a treatment of high-quality speech transmission by AM and FM.

In Fig. 5–6 the input ratios for each system are shown as separate abscissae to allow a quick comparison on an equal signal power basis. The relative displacements of the x-axes, determined by

$$\frac{S}{N}\bigg|_{\substack{\text{in} \\ \text{full AM}}} = \frac{S}{N}\bigg|_{\substack{\text{in} \\ \text{DSBSC}}} = \tfrac{1}{2}\frac{S}{N}\bigg|_{\substack{\text{in} \\ \text{SSB}}} = \frac{B_{\text{IF}(\text{FM})}}{B_{\text{IF}(\text{AM})}}\frac{S}{N}\bigg|_{\substack{\text{in} \\ \text{FM}}}$$

are arranged so that a line drawn parallel to the y-axis represents equal power for all systems. The output signal-to-noise ratios are calculated for systems handling a nominal 3 kHz baseband signal.

Curve I—full AM with envelope detection—represents Eq. (5–11) with $m_a = 1$, i.e.

$$\frac{S}{N}\bigg|_{\substack{\text{out} \\ \text{env. det.}}} \approx \frac{1}{1\cdot5}\frac{S}{N}\bigg|_{\substack{\text{in} \\ \text{full AM}}}$$

$\equiv 1\cdot75$ dB reduction in signal-to-noise power ratio.

Note that this is only valid for high signal-to-noise ratios.

Curve II—full AM with square-law detection—represents Eq. (5–24) with $m_a = 1$, i.e.

$$\frac{S}{N}\bigg|_{\substack{\text{out} \\ \text{sq. law det.}}} \approx \frac{1}{2\cdot25}\frac{S}{N}\bigg|_{\substack{\text{in} \\ \text{full AM}}}$$

$\equiv 3\cdot5$ dB reduction in signal-to-noise power ratio.

Curve III represents both DSBSC and SSB systems. It was shown in section 5–3 that the output signal-to-noise ratios are the same when both systems have equal signal power. The curve is therefore a plot of Eq. (5–27) with $\phi_c = \phi_{osc}$, and Eq. (5–31).

Curve IV is plotted for an FM system having a peak frequency deviation $\Delta f = 3\cdot6$ kHz and a channel bandwidth $B_{\text{IF}} = 12\cdot5$ kHz. Inserting these figures in Eq. (5–45a) gives

[5–6]

$$\left.\frac{S}{N}\right|_{\substack{\text{out}\\\text{FM}}} \approx 3\left[\frac{3\cdot6}{3}\right]^2 \frac{12\cdot5}{2\times3} \left.\frac{S}{N}\right|_{\substack{\text{in}\\\text{FM}}} = 9\left.\frac{S}{N}\right|_{\substack{\text{in}\\\text{FM}}}$$

$\equiv$ 9·5 dB improvement in the signal-to-noise ratio.

Using Eq. (5–47a), the FM improvement over a full AM system using envelope detection is approximately 8 dB.

Curve V is plotted for a FM system having a peak frequency deviation $\Delta f = 6$ kHz and a channel bandwidth $B_{IF} = 25$ kHz. Eq. (5–45a) becomes

$$\left.\frac{S}{N}\right|_{\substack{\text{out}\\\text{FM}}} \approx 3\left[\frac{6}{3}\right]^2 \frac{25}{2\times3} \left.\frac{S}{N}\right|_{\substack{\text{in}\\\text{FM}}} = 50\left.\frac{S}{N}\right|_{\substack{\text{in}\\\text{FM}}}$$

$\equiv$ a 17 dB improvement in the signal-to-noise ratio.

Using Eq. (5–47a), the FM improvement over a full AM system using envelope detection is approximately 12·5 dB.

An inspection of Fig. 5–6 shows that when the signal-to-noise ratio is high there is a considerable advantage to be gained from using an FM system and the improvement in dB over an AM system increases in direct proportion to the peak frequency deviation. But for a given signal power there is a limit to the peak frequency deviation for which the full FM improvement can be realised; note from Curves IV and V that the wider band system has a lower input signal-to-noise ratio and the second threshold is further to the right (see also Tutorial Exercise 5–2). At low signal-to-noise power ratios it can be seen that an SSB system is superior to both FM and non-coherent AM systems.

Tutorial Exercises

5–1 What is meant by *coherent detection* of an amplitude modulated signal? Explain the difference between detecting a double sideband suppressed carrier (DSBSC) signal and a single sideband (SSB) signal.

The signal-to-noise ratio of a DSBSC signal in the presence of band-limited uniform spectral density noise is 15 dB. A coherent detector followed by a filter is used to recover the baseband signal. Calculate the maximum signal-to-noise ratio of the output.

Ans: 18 dB.

5–2 The reception of an unmodulated carrier is accompanied by an interfering signal 22 dB below the level of the required carrier and with a frequency separation of 10 kHz. The demodulator of the receiver is a frequency-conscious discriminator preceded by an amplitude limiter; the receiver is correctly tuned to the frequency of the unmodulated carrier.

Calculate the peak amplitude and frequency of the demodulated output due

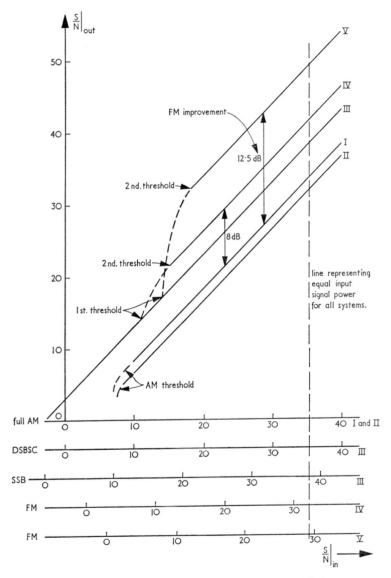

Fig. 5–6 Demodulation characteristics of various analogue modulation systems.

to the interfering signal. Assume that the discriminator output voltage/frequency deviation characteristic is linear with a slope of 100 mV/kHz and that the interfering signal falls within the passband of the receiver.

Ans: 79·4 mV; 10 kHz.

5–3 An SSB signal in the presence of band-limited Gaussian-distributed noise is applied to the input of a chopper demodulator. Show that the signal-to-noise power ratio for the filtered baseband signal is equal to the input signal-to-noise ratio.

5–4 What are the advantages of using wideband frequency modulation for the transmission of analogue signals?

The maximum range of a high-quality sound broadcast transmission, contained in a 15 kHz baseband, is set by the minimum acceptable signal-to-noise ratio at the receiver output. This minimum is 60 dB when using full AM with sine wave modulation and a modulation index $m_a = 0.8$.

Calculate the theoretical increase in the transmission range for the same receiver output signal-to-noise ratio if frequency modulation with a peak frequency deviation of 60 kHz is used.

Assume

(i) the transmitter power is the same for both systems;
(ii) the signal power decreases according to the square of the transmission distance, and
(iii) the noise has uniform spectral density.

Ans: factor of 10.

5–5 In a full AM system handling a 3 kHz baseband signal the input and output signal-to-noise power ratios of the envelope detector are 20 and 18 dB respectively.

In an FM system which is to carry the same information, determine the maximum peak frequency deviation for which full FM improvement can be realised and also calculate this improvement. Base the calculations on single sine wave modulation at the highest baseband frequency. Also assume

(i) that the signal-to-noise ratio at the input to the FM demodulator and corresponding to the threshold level is 12 dB,
(ii) that both systems have equal signal power at the inputs of the respective demodulators, and
(iii) the significant sideband criterion defined in Chapter 2.

Ans: 11·2 kHz; 18 dB.

6
PULSE ANALOGUE MODULATION

6–1 Sampling

In the preceding chapters the modulating signal $m(t)$ has been considered as a continuously varying amplitude-time function. Let us now consider this signal to be present for only part of the time as a series of pulses with abrupt discontinuities between each pulse interval, as shown in Fig. 6–1 (*b*). If this is the received demodulated output corresponding to the modulating signal $m(t)$, it follows intuitively that it should be

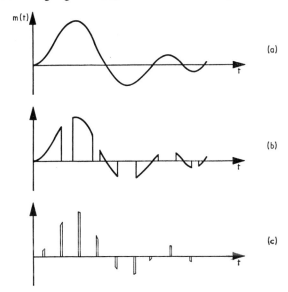

Fig. 6–1 (*a*) Continuous modulating signal; (*b*) discontinuous version, and (*c*) signal represented as a series of pseudo-impulses.

possible to reconstruct faithfully the original modulating signal—it is assumed that there are no abrupt changes in $m(t)$ during the intervals between pulses and furthermore the signal is present for a significantly large fraction of the total time. We therefore conclude that it is not necessary to transmit $m(t)$ for all of the time.

[6–1]

Another discontinuous version of the modulating signal, shown in Fig. 6–1 (*c*), consists of a series of pulses (or pseudo impulses; see Appendix I, page 233) with duration t_p, which is small compared with the reciprocal of the highest baseband frequency. The amplitude of the modulating signal is virtually constant over the pulse interval and therefore the pulse may be considered to represent the instantaneous amplitude of $m(t)$. A common method of implementation is a sampling gate[1] supplied with the modulating signal and a stream of constant-amplitude

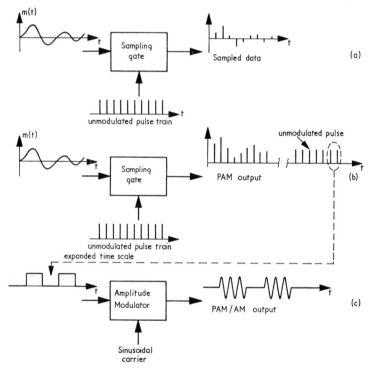

Fig. 6–2 Generation of (*a*) sampled data; (*b*) PAM, and (*c*) PAM/AM.

pulses (see Fig. 6–2 (*a*)). The output, which is referred to as *sampled data*, is zero when $m(t)$ is zero, and therefore if we think of the constant-amplitude pulse stream as a carrier the output is a pulse amplitude modulated signal with suppressed carrier. An alternative arrangement produces an output consisting of constant amplitude pulses, i.e. the carrier, when the input signal $m(t)$ is zero—this is known as *pulse amplitude modulation* (PAM) (see Fig. 6–2 (*b*)), as distinct from sampled data.

1. *Pulse, Digital and Switching Waveforms*, J. Millman and H. Taub, Chapter 17 (McGraw-Hill, 1965).

Sampled data and PAM signals are essentially baseband signals, although they occupy a much wider bandwidth than their equivalent continuous modulating signal. Translation to radio frequencies can be accomplished by any of the conventional forms of modulation described previously. For example, if a PAM signal amplitude modulates a sinusoidal carrier in an on-off mode, the carrier has a peak amplitude proportional to the amplitude of the pulse and is only present during the pulse interval. Each pulse will give rise to many cycles of the carrier, as the condition $1/f_m \gg t_p \gg 1/f_c$ can be established with little difficulty—f_m represents the highest baseband frequency and f_c the sinusoidal carrier frequency; the RF output is referred to as a PAM/AM signal. Alternatively, frequency modulation could be used to generate a PAM/FM signal in which the frequency deviation of the carrier is determined by

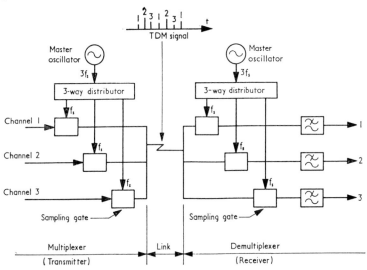

Fig. 6–3 Multiplexing and demultiplexing of a 3-channel TDM system.

the amplitude of the modulating pulse. Apart from this brief reference to pulse modulation of a carrier the following analysis is concerned solely with the baseband signal.

Having illustrated that the analogue signal need not be transmitted in its entirety, we may now introduce the important principle of *time division multiplex* (TDM). It has been shown in earlier chapters that a number of independent messages can be accommodated in a communication system by sharing the allocated bandwidth, i.e. *frequency division multiplex* (FDM). On the other hand, in a TDM system the

[6–1]

total bandwidth is available to all channels, but the time is shared between them. Fig. 6–3 illustrates a method of multiplexing three channels in which the sampling gates are operated sequentially to produce a combined output of interleaved-channel pulses. The question to be posed is 'How many channels can be accommodated in such a system or, in other words, how frequently must each channel be sampled to allow faithful reproduction of the message at the receiving terminal?' To answer this question let us return to the sampling of a single channel, as depicted in Fig. 6–2. The unmodulated carrier of constant-amplitude pulses, at a rate f_s per second, has a line spectrum as shown in Fig. 6–4 (*a*). It is assumed that the pulses are effectively pseudo-impulses satisfying the condition $t_p \ll 1/f_m$ and consequently the spectral lines

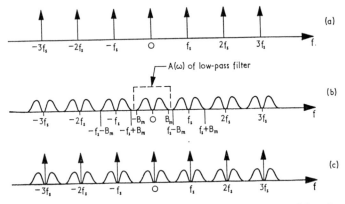

Fig. 6–4 Spectrum of (*a*) constant amplitude periodic sequence of impulses; (*b*) sampled data, and (*c*) a PAM signal.

are approximately of equal amplitude over the frequency range to be considered. If the continuous modulating signal is band-limited, the sampled data and PAM spectra are as shown in Figs. 6–4 (*b*) and 6–4 (*c*) respectively; it is assumed that the baseband is B_m and, for convenience of presentation, that this is equal to the highest modulating frequency, i.e. the baseband extends to zero frequency. Strictly speaking, this need not be the case, as we have seen for speech the baseband is essentially 300–3400 Hz. It is to be noted that the spectra are shown for $B_m < f_s/2$ and consequently there is no overlapping of sidebands associated with adjacent spectral lines. This means that, if either the sampled data or the PAM signal is applied to a low-pass filter capable of extracting only those components within the baseband B_m, the original continuous signal is recoverable. But with $B_m > f_s/2$ the lower sideband associated with the first spectral line will overlap the baseband spectrum and the filter output would be a distorted version of the original continuous signal. We therefore conclude that the pulse rate f_s must be equal to or greater than twice the highest frequency of the baseband signal for undistorted reconstitution of the continuous signal at a receiving terminal. Referring back to the three-channel system of Fig. 6–3 and the question that it posed, we see that maximum capacity is realised when the pulse generator supplying the three-way distributor is operating at a rate $3 \times 2B_m$ pulses per second; for example, when the three inputs are speech signals, the minimum overall sampling rate is $3 \times 2 \times 3400 = 20400$ pulses per second. In practice perfect band-limiting cannot be achieved and a compromise must be reached between an acceptable level of distortion of the recovered signal and a sampling rate that is in excess of the theoretical minimum of $2B_m$ pulses per second. This is dealt with later in the chapter under *Aliasing*.

A mathematical treatment of the sampling of a band-limited signal has been established by Shannon and is universally known as Shannon's Sampling Theorem.[2] Although the above explanation based on the

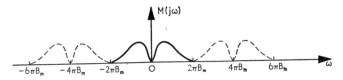

Fig. 6–5 Band-limited spectrum of modulating signal $m(t)$.

spectral distribution of the sampled signal will suffice for our requirements, an account of the Sampling Theorem is given below, since it is a classic example of the use of Fourier series and Fourier integral techniques. The theorem states that, if $f(t)$ is a band-limited function, it is completely defined by giving its ordinates at a series of points $1/2B_m$ seconds apart, where B_m is the highest frequency of the function and usually equal to the bandwidth.

Let us consider that the modulating signal exists for a finite time T, i.e. it is a truncated signal which we shall refer to as $m_T(t)$. Using the Fourier integral,

$$m_T(t) = \int_{-B_m}^{B_m} M(j\omega)\exp(j\omega t)df \qquad (6\text{–}1)$$

where $M(j\omega)$ defines the amplitude spectral density of $m_T(t)$.

2. Communication in the Presence of Noise, C. E. Shannon, *Proc. I.R.E.*, Vol. 37, No. 1, pp. 10–21, January 1949.

[6–1]

If we assume that the band-limited spectrum is one complete period of a periodic function defined in the *frequency domain*, $M(j\omega)$ may be represented as a Fourier series,

$$M(j\omega) = \sum_{n=-\infty}^{\infty} x_n \exp\left(j\frac{2\pi nf}{2B_m}\right), \text{ for } -B_m < f < B_m \quad (6\text{-}2)$$

where

$$x_n = \frac{1}{2B_m} \int_{-B_m}^{B_m} M(j\omega) \exp\left(-j\frac{2\pi nf}{2B_m}\right) df \quad (6\text{-}3)$$

It is interesting to compare this representation of a periodic function defining a frequency-dependent variable with the more familiar time domain series of Appendix I, page 212. Comparing Eqs. (6-2) and (I-8), we see that $f \equiv t$ and $2B_m \equiv T$.

The sampling of $m_T(t)$ at regular intervals means that

$$m_T(t) \equiv m\left(\frac{n}{f_s}\right) \text{ or } m\left(-\frac{n}{f_s}\right)$$

where f_s is the sampling rate and n is any positive or negative integer. Therefore replacing t by $-n/f_s$ in Eq. (6-1) gives

$$m\left(-\frac{n}{f_s}\right) = \int_{-B_m}^{B_m} M(j\omega) \exp\left(-j\frac{2\pi nf}{f_s}\right) df \quad (6\text{-}4)$$

For $f_s = 2B_m$, we see from a comparison of Eqs. (6-3) and (6-4) that

$$m\left(-\frac{n}{2B_m}\right) = 2B_m x_n$$

i.e. samples at . . . $-3/2B_m$, $-2/2B_m$, $-1/2B_m$, 0, $1/2B_m$, $2/2B_m$, $3/2B_m$ . . . define the coefficient x_n which in turn defines $F(j\omega)$. If $f_s > 2B_m$ this is equivalent to increasing the number of components defining x_n, but when $f_s < 2B_m$ there are insufficient samples to define the coefficient.

6-2 Reconstitution of the Original Signal

A continuous signal can be formed from the demultiplexed pulse train by an interpolation process that fills the gaps between the pulses. If the reconstituted signal is $u(t)$, say, then

$$u(t) = \sum_n g_n h\left(t - \frac{n}{f_s}\right) \quad (6\text{-}5)$$

where g_n defines the amplitude of the nth pulse and $h(t - n/f_s)$ is the interpolation function corresponding to g_n. The process is best achieved by applying the pulses—which, remember, are effectively pseudo-impulses, since $t_p \ll 1/B_m$—to a network having an impulse response equal to the required interpolation function. The following examples illustrate the principle of reconstitution and also enable us to deduce the interpolation function which yields an output $u(t)$ that is identical to the required signal $m_T(t)$.

6–2.1 *Staircase Interpolation*
It can be seen from Fig. 6–6 that the interpolation function is rectangular with a duration equal to the sampling interval $1/f_s$. Since this function

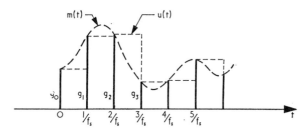

Fig. 6–6 Staircase interpolation.

must be the impulse response of the network to which the pulse train is applied, the network transfer function is given by the Fourier integral

$$H(j\omega) = \int_0^{1/f_s} \exp(-j\omega t)dt$$

$$= \frac{\sin(\omega/2f_s)}{\omega/2f_s} \exp(-j\omega/2f_s)$$

where the limits of integration satisfy the interpolation process over the interval $t = 0$ to $t = 1/f_s$.

The $\sin x/x$ filter response can be realised by a relatively simple sample-hold circuit.[3] The pulse charges a capacitor from a low-impedance source which is then disconnected electrically to enable the capacitor to hold the charge over the time interval $1/f_s$. At the next sampling instant the capacitor assumes a new value of charge corresponding to the magnitude of the particular input pulse. There is always an exponential decay of the stored charge, which can be minimised by having as high a load impedance as possible—MOSFETs are particularly suitable in this application.

3. *Electronic Analogue and Hybrid Computers*, G. A. Korn and T. M. Korn, Chapter 10 (McGraw-Hill, 1964).

[6–2]

The marked discrepancy between the original and reconstituted signals, $m(t)$ and $u(t)$ respectively, is clearly seen in Fig. 6–6. The next method to be described gives a closer approximation.

6–2.2 *Linear Interpolation*

The interpolation function is triangular and extends over a time interval $2/f_s$. For example, at $t = 0$, $n = 0$ and $g_n = g_0$, the interpolation function is defined as

$$h(t-n/f_s) = h(t) = f_s t \text{ for } 0 \leqslant t \leqslant 1/f_s$$

and
$$h(t) = (2-f_s t) \text{ for } 1/f_s \leqslant t \leqslant 2/f_s$$

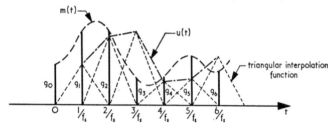

Fig. 6–7 Linear interpolation.

The transfer function of the required network, triangular impulse-response, is given by

$$H(j\omega) = \int_0^{1/f_s} f_s t \exp(-j\omega t)dt + \int_{1/f_s}^{2/f_s} (2-f_s t)\exp(-j\omega t)dt$$

$$= \frac{1}{f_s} \frac{\sin^2(\omega/2f_s)}{(\omega/2f_s)^2} \exp(-j\omega/f_s)$$

(see also the Tutorial Exercise on page 222).

This transfer function can be realised by using two $\sin x/x$ filters in tandem. A suitable arrangement is shown in Fig. 6–8—it is left to the

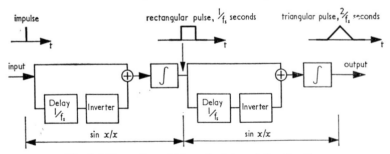

Fig. 6–8 Generation of triangular interpolation function.

reader as an exercise to show that the impulse response is triangular. It will be appreciated that the sample-hold circuit described previously does not lend itself to this application.

6–2.3 *Ideal Interpolation*

A comparison of Figs. 6–6 and 6–7 shows that linear interpolation gives a closer approximation to the original signal. We also note that for staircase interpolation $h(t-n/f_s)$ is piecewise continuous and gives rise to abrupt changes in $u(t)$, whereas the linear interpolation function is continuous and abrupt discontinuities are found only in the first derivative. In fact, it can be shown, although of academic interest only, that the interpolation function must have continuous derivatives of higher orders if $u(t)$ is to resemble closely $m_T(t)$. It therefore follows that a band-limited interpolation function is required, as its derivatives of all orders are continuous, i.e. the smoothest interpolation function is band-limited.

To arrive at the required function let us consider the original band-limited signal $m_T(t)$ defined by

$$m_T(t) = \int_{-B_m}^{B_m} M(j\omega)\exp(j\omega t)df$$

Also from the proof of the Sampling Theorem,

$$M(j\omega) = \sum_{n=-\infty}^{\infty} x_n\exp\left(j\frac{n\pi f}{B_m}\right)$$

Therefore, $m_T(t) = \int_{-B_m}^{B_m} \sum_{n=-\infty}^{\infty} x_n\exp\left(j\frac{n\pi f}{B_m}\right)\exp(j\omega t)df$

$$= \sum_{n=-\infty}^{\infty} x_n \int_{-B_m}^{B_m} \exp\left[jf\left(2\pi t+\frac{n\pi}{B_m}\right)\right]df$$

$$= \sum_{n=-\infty}^{\infty} x_n\, 2B_m\, \frac{\sin[2\pi B_m(t+n/2B_m)]}{2\pi B_m(t+n/2B_m)} \qquad (6\text{-}6)$$

But if $m_T(t)$ is represented as a sampled signal and Shannon's Sampling Theorem is satisfied,

$$m\left(-\frac{n}{f_s}\right) = 2B_m x_n$$

and therefore substituting for x_n in Eq. (6–6) gives

$$m_T(t) = \sum_{n=-\infty}^{\infty} m\left(-\frac{n}{2B_m}\right) \frac{\sin[2\pi B_m(t+n/2B_m)]}{2\pi B_m(t+n/2B_m)} \qquad (6\text{–}7)$$

Comparing this equation with Eq. (6–5) gives

$$g_n \equiv m\left(-\frac{n}{2B_m}\right)$$

and therefore $u(t)$ and $m_T(t)$ are identical when the interpolation function

$$h\left(t-\frac{n}{f_s}\right) = \frac{\sin[2\pi B_m(t+n/2B_m)]}{2\pi B_m(t+n/2B_m)}$$

which, of course, is the output response of an ideal low-pass filter (cut-off frequency $f_o = B_m$) to an impulse occurring at time $t = n/f_s$, where $f_s = 2B_m$. For example, when $n = 0$ and the input is a pseudo-impulse of unit amplitude and duration t_p, the output response of the filter is

$$h(t) = \int_{-B_m}^{B_m} 1 \times t_p A \exp(-j\omega T_d)\exp(j\omega t)df$$

$$= 2t_p \, A \, B_m \, \frac{\sin[2\pi B_m(t-T_d)]}{2\pi B_m(t-T_d)}$$

$$= const. \, \frac{\sin[2\pi B_m(t-T_d)]}{2\pi B_m(t-T_d)}$$

where $A \exp(-j\omega T_d)$ is the filter transfer function (see page 225). For the general case

$$h\left(t-\frac{n}{2B_m}\right) = const. \, \frac{\sin[2\pi B_m(t-T_d+n/2B_m)]}{2\pi B_m(t-T_d+n/2B_m)}$$

When the input to the filter is a train of amplitude-modulated pulses, the interpolation functions are weighted by amplitudes of the pulses to give the required output

$$u(t) = \sum_{n} g_n \, const. \, \frac{\sin[2\pi B_m(t-T_d+n/2B_m)]}{2\pi B_m(t-T_d+n/2B_m)} \qquad (6\text{–}8)$$

The linear phase-shift characteristic and perfect cut-off assumed for the filter makes the method physically unrealizable, as can also be seen from the precursor associated with the interpolation function response in Fig. 6–9. Nevertheless, it does point to the practical requirement of a filter having as near a perfect cut-off as possible for the reconstitution

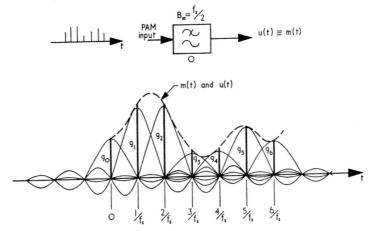

Fig. 6–9 Ideal interpolation.

of the original signal; the unavoidable non-linear phase-frequency characteristic of a realizable filter removes the problem of a precursor (see also Appendix I, page 229).

6–3 Aliasing

So far we have assumed perfect band-limiting of the modulating signal and considered the sampling rate to be at least twice the highest frequency of this signal. In practice band-limiting is always imperfect and

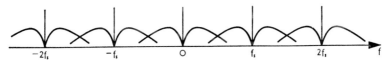

Fig. 6–10 Part of the spectrum of a PAM signal corresponding to an imperfectly band-limited input.

consequently the condition stipulated by the Sampling Theorem is violated. This is best illustrated by the spectral representation of sampled data (see Fig. 6–10 and compare it with Fig. 6–4). It is seen that the sidebands associated with adjacent spectral lines overlap and in particular part of the LSB of the first spectral line will be indistinguishable

[6–3]

from the original baseband spectrum when reconstituting the continuous signal by low-pass filtering. The resulting distortion (or *aliasing*) has been studied by Bode and Shannon[4] and their method of analysis is illustrated in Fig. 6–11. They have shown that the aliasing error power,

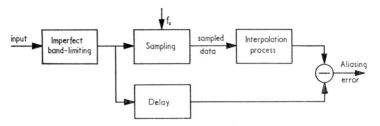

Fig. 6–11 Block diagram illustrating the Bode and Shannon method for determining the optimum interpolation process for minimum aliasing error.

which is evident at the higher baseband frequencies, can be minimised by employing a de-emphasising characteristic in the interpolation process. It is felt that the analysis is outside the scope of an introductory text and therefore a simpler approximate method of deducing the aliasing error is given instead.

Consider a practical filter having a Butterworth response[5, 6] defined by the transfer impedance

$$\left| Z_{21} \right| = \left| \frac{v_{\text{out}}}{i_{\text{in}}} \right| = \frac{1}{\sqrt{1+(f/f_0)^{2m}}}$$

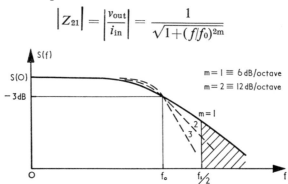

Fig. 6–12 Spectral noise power density of the output from a filter having a Butterworth response to which uniform spectral density noise is applied.

where f_0 is the 3 dB break-point frequency and m determines the slope of the cut-off characteristic (see Fig. 6–12). For a 1 ohm load, the

4. A Simplified Derivation of Linear Least Square Smoothing and Prediction, H. W. Bode and C. E. Shannon, *Proc. I.R.E.*, Vol. 38, No. 4, pp. 417–25, April 1950.
5. *Linear Electric Circuits*, W. L. Cassell, Chapter 20 (Wiley, 1964).
6. *Introductory Circuit Theory*, E. A. Guilleman, Chapter 9 (Wiley, 1953).

normalised power $P_{out}(f)$ is given by

$$P_{out}(f) = \overline{i_{in}^2}/[1+(f/f_0)^{2m}]$$

and at very low frequencies, i.e. $f \to 0$, $P_{out}(0) = \overline{i_{in}^2}$

and therefore $P_{out}(f) = P_{out}(0)/[1+(f/f_0)^{2m}]$

for constant input current. If the signal that is to be band-limited has a uniform spectral power density, the output spectral density from the filter is given by

$$S(f) = \frac{S(0)}{1+(f/f_0)^{2m}} \quad \text{watts/Hz} \qquad (6\text{–}9)$$

(see Fig. 6–12).

The aliasing distortion may be assessed by evaluating the total energy of the signal and the energy contained in the shaded area which will undergo a downward spectral transposition in the reconstitution process, since the Sampling Theorem is not obeyed. The aliasing error power is given by

$$S_{aliasing} = \int_{f_s/2}^{\infty} \frac{S(0)}{1+(f/f_0)^{2m}} \, df$$

The integrand is simplified by assuming $1+(f/f_0)^{2m} \approx (f/f_0)^{2m}$, which is a reasonable approximation, since when $m = 1$ a large proportion of the aliasing error power is due to components having frequencies considerably in excess of f_0, and when $m \geqslant 2$, $(f/f_0)^{2m} \gg 1$. Therefore,

$$S_{aliasing} \approx \int_{f_s/2}^{\infty} S(0) \left(\frac{f_0}{f}\right)^{2m} df = \frac{S(0)f_0 2^{2m-1}}{2m-1} \left(\frac{f_0}{f_s}\right)^{2m-1} \quad \text{watts} \quad (6\text{–}10)$$

The total signal power $S_T = \int_0^{\infty} \frac{S(0)}{1+(f/f_0)^{2m}} \, df$

The definite integral $\int_0^{\infty} \frac{dx}{1+(x/a)^n} = \frac{\pi a}{n} \operatorname{cosec}\left(\frac{\pi}{n}\right)$, and therefore

$$S_T = \frac{S(0)\pi f_0}{2m} \operatorname{cosec}\left(\frac{\pi}{2m}\right) \qquad (6\text{–}11)$$

The r.m.s. aliasing error can be expressed as a percentage of the total signal power, viz.

$$\left[\frac{S_{aliasing}}{S_T}\right]^{\frac{1}{2}} = 2m \left[\frac{m}{\pi(2m-1)} \left(\frac{f_0}{f_s}\right)^{2m-1} \sin\left(\frac{\pi}{2m}\right)\right]^{\frac{1}{2}} \times 100 \qquad (6\text{–}12)$$

$$[6\text{–}3]$$

and in Fig. 6–13 it is plotted against the normalised sampling frequency f_s/f_0 for various values of m. We see that for a given aliasing error a compromise must be made between a low sampling rate and a complicated filter. For speech applications the sampling rate is usually 8000 p.p.s. and, if the aliasing error is to be 26 dB below the mean signal level,

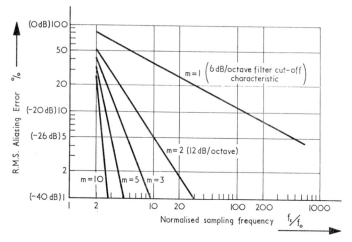

Fig. 6–13 Aliasing error *v.* normalised sampling frequency in terms of the Butterworth filter cut-off characteristic.

the band-limiting filter must have a 40–50 dB/octave cut-off characteristic.

6–4 Effect of Band-limiting

A periodic train of rectangular pulses occupies, theoretically, the entire frequency spectrum and, in practice, would occupy an extremely wide band if the rise and decay times and rounding off of the pulses are negligible. The communication system, including the transmission medium, should have ideally a bandwidth sufficient to accommodate the pulse spectrum; however, it has been discussed previously that the signal bandwidth must, in the case of radio transmission, be restricted in order to avoid interference to other services using the same medium. With line systems, although it is not essential to restrict the bandwidth of the signal prior to transmission, the spectrum is unavoidably curtailed *en route*, since the cable and repeaters have a limited spectral capacity. We have, therefore, in both radio and line systems, signals of restricted bandwidth and if the original TDM pulses are rectangular— or any other shape provided that the energy extends over an appreciably wider band than that available in the overall communication system

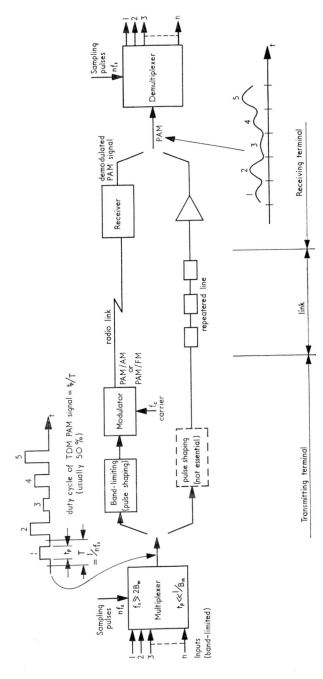

Fig. 6-14 Radio and line transmission systems for TDM analogue signals.

the baseband signals at the receiver are made up of pulses that overlap each other to give *intersymbol* (or inter-pulse) *interference* (see Fig. 6–14).

Let us consider this phenomenon for line transmission. The high-frequency cut-off characteristic of the cable and repeaters can be simulated by a simple *RC* network (see Fig. 6–15). The time interval between adjacent pulses is t_g and the overlap time is Δ, where $t_g+\Delta$

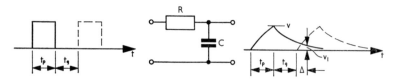

Fig. 6–15 Illustration of pulse distortion and intersymbol interference.

is the time taken for the pulse amplitude to decay from V to V_1; consequently

$$V_1 = V \exp \{-(t_g+\Delta)/RC\}$$

and the *crosstalk ratio* is by definition V/V_1, which expressed in dB is $8{\cdot}686\,[(t_g+\Delta)/RC]$ dB. Restating in terms of the 3 dB upper cut-off frequency of the system, we have

$$f_{3\,dB} = \frac{1}{2\pi RC} = \frac{\text{crosstalk ratio (in dB)}}{2\pi \times 8{\cdot}686(t_g+\Delta)} \text{ Hz} \qquad (6\text{–}13)$$

An an example, consider the time multiplexing of twenty-four 3-kHz channels at an overall sampling rate of $24 \times 8000 = 192 \times 10^3$ p.p.s. to produce a PAM signal having a 50 % duty cycle, i.e. $t_p = t_g = 2{\cdot}6\mu s$. Usually a crosstalk ratio of 60 dB is desirable and if we assume that this ratio must be realised at the commencement of the pulses, $\Delta = 0$, then according to Eq. (6–13),

$$f_{3\,dB} = \frac{60}{2\pi \times 8{\cdot}686 \times 2{\cdot}6 \times 10^{-6}} = 420 \text{ kHz}$$

It is interesting to compare this figure with the bandwidth of 96 kHz $(24 \times 4 \text{ kHz})$ required by a FDM system. The considerably greater bandwidth required for the analogue TDM system is one of the major drawbacks to its use. Theoretically, this need not be the case; it has been shown by Nyquist[7] that the minimum transmission bandwidth for the TDM signal, for which crosstalk can be avoided completely, is numerically half the pulse rate—in the above example this would be $192 \times 10^3/2 = 96$ kHz. In order to achieve this ideal condition the PAM signal must

7. Certain Topics in Telegraph Transmission Theory, H. Nyquist, *Trans. A.I.E.E.*, Vol. 47, pp. 617–44, April 1928.

be made up of sin x/x pulses which can be generated by applying impulses at a rate of f_s p.p.s. to an ideal low-pass filter with cut-off frequency $f_s/2$ Hz, similar to the ideal interpolation process illustrated in Fig. 6–9 (see also section I–6, page 232). It can be seen that there is considerable overlapping of the pulses, but at the instant when a pulse is at its maximum amplitude all other pulses are zero. Therefore, if the receiver demultiplexer is arranged to inspect the incoming signal every $1/f_s$ seconds and the inspection times are synchronised with the occurrence of the maximum pulse amplitudes, the time-interleaved channels can be separated with zero crosstalk. In practice, although an extremely close approximation to the sin x/x pulses could be realised, this seemingly attractive arrangement is not used, simply because the transmitter pulse rate, filter cut-off frequency and receiver inspection rate would have to be held to prohibitively close tolerances, otherwise the overlapping pulse tails could add up to very large amplitudes and so produce considerable crosstalk. Whilst it is possible to generate band-limited pulses of similar shape[8] to the sin x/x pulse, but with more rapidly converging tails, they have not found application for pulse analogue transmission, again due to the constraints that would have to be placed on the tolerances of the system parameters: a detailed description of these pulses and their application are given in Chapter 8.

We shall not consider the transmission of TDM signals in analogue form in any further detail since nowadays digital data methods are used almost exclusively. It must be stressed, however, that the sampling of the continuous message to form a sampled data or PAM signal, covered in sections 6–1 to 6–3, applies equally to analogue and digital transmission systems. In the following chapters it will be seen that there are many advantages to be gained from converting the analogue pulses into a digital signal prior to transmission. One is that intersymbol interference is less troublesome; the receiver is not concerned with detecting an infinite number of amplitude levels within the dynamic range of the signal but only in deciding the nearest level to a particular input amplitude out of a set of two for a binary system or 3, 4, etc. for a m-ary system. Another distinct advantage is that the receiver decision process has the inherent capability of alleviating noise; once the receiver has made the decision regarding a particular input amplitude and the nearest level, provided that it is the correct decision, the noise is virtually removed from the signal. Whilst this brief reference to digital methods of transmission is intended to justify the omission of a detailed account of

8. Theoretical Fundamentals of Pulse Transmission, E. D. Sunde, *Bell System Tech. J.*, Vol. 33, pp. 721–88, May 1954.

[6–4]

pulse analogue transmission systems, it also serves as an introduction to the following chapters.

Tutorial Exercises

6–1 A train of constant amplitude rectangular pulses is defined by the Fourier series,

$$f(t) = 5\left\{ \frac{1}{10} + \sum_{n=1}^{\infty} \frac{2}{n\pi} \sin\left(\frac{n\pi}{10}\right) \cos\left[2\pi n \times (5 \times 10^3)t\right] \right\} \quad \text{volts}$$

The pulse train is used as a carrier and amplitude modulated to a depth of 80% by a 1·5 kHz sine wave. If the resulting signal is passed through a 4 kHz ideal band-pass filter centred on 25 kHz, calculate the r.m.s. output voltage from the filter.

Ans: 0·255 volts.

6–2 A continuous signal which is band-limited by a maximally flat filter, has a power spectral density defined by

$$S(f) = \frac{1}{1 + (f/f_0)^{20}} \quad \text{watts/Hz}$$

where the half power 3 dB break frequency $f_0 = 3·4$ kHz.

The signal is sampled 8000 times per second and transmitted in PAM form. Calculate, approximately, the r.m.s. aliasing error when the signal is reconstituted.

Ans: 4·8%.

6–3 A signal occupies a bandwidth from d.c. up to the very high audio frequencies. For a certain application, it is necessary to transmit, by pulse modulation, the information contained in part of this spectrum.

The signal is band-limited by two parallel RC networks (see Fig. TE 6–1), and sampled 3000 times per second. Calculate, by an approximate method, the error due to aliasing when the continuous signal is reconstructed.

Ans: 0·33%.

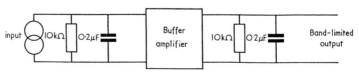

Fig. TE 6–1.

7
ANALOGUE-TO-DIGITAL CONVERSION (QUANTIZATION)

The reasons for converting analogue signals into digital form prior to transmission have been outlined in the concluding remarks of Chapter 6, and a more detailed explanation is to be found in Chapter 8. However, first let us look at ways of representing the information digitally; there are in general two possibilities.

One is known as *pulse-code modulation* (PCM), in which the amplitude range of the sampled data is divided into a finite number of discrete levels; the amplitude of a given pulse is referred to the nearest level and a digital code generated—for example, in a binary system 31 amplitude levels (or quanta) are uniquely specified by a 5-digit code, since $2^5-1 = 31$. The analogue-to-digital conversion introduces an irremovable error known as *quantization noise* which depends upon the number of levels and hence upon the number of digits in the code—the larger the number of levels the smaller the noise. It will be seen that for a given number of code digits, which is determined by the available bandwidth, the quantization distortion can be minimised by choosing nonuniform spacing of the levels to suit the statistical properties of the signal.

The second method of conversion is known as *delta modulation* (ΔM) or differential PCM and makes use of a 1-digit code. In the basic system, as opposed to delta-sigma modulation ($\Delta-\Sigma$M) which is also described, the derivative of the input is transmitted rather than the instantaneous amplitude as in PCM. This is achieved by integrating the digitally encoded signal and comparing it with the analogue input in order to decide which of the two has the larger amplitude; the polarity of the next pulse, either $+V$ or $-V$, is chosen to reduce the difference in amplitude between the two waveforms. The receiver need only be an integrator followed by a low-pass filter, which removes the abrupt step-like characteristics of the waveform, to produce an output that closely follows the original analogue input. Such a process of successive approximations must introduce an error similar to that of quantization noise in PCM. The evaluation of this noise by a rigorous analysis is

[7]

extremely complex and is not undertaken here, but instead the results of recent numerical analyses are assumed and have been found to be in reasonable agreement with previous methods of assessing the signal-to-quantizing noise ratio. Companded delta modulation is discussed and the conclusion reached is that this system offers a comparable specification to PCM for transmitting speech and has the advantage of simplicity of circuitry.

7–1 Pulse Code Modulation (PCM)

7–1.1 *Quantization Noise*

Misrepresentation cannot be avoided in the conversion of the signal from an analogue to a digital form. It can be seen from Fig. 7–1 that the sampled data pulses are referred to the nearest quantization level in order that a binary code may be generated; in this illustration there are 15 levels giving rise to a 4-digit binary code.

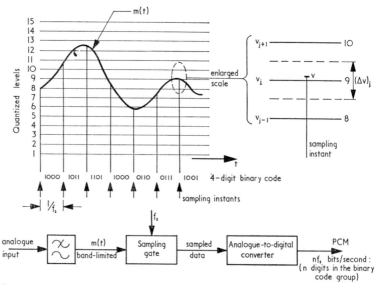

Fig. 7–1 Quantization of an analogue signal and the basic arrangement for generating PCM.

Consider the instantaneous amplitude of the analogue signal to be v and the nearest quantizing level v_j. The quantizing error will be due to $(v-v_j)$ and the mean square error voltage is $\overline{(v-v_j)^2}$. All the signal amplitudes within the range $v_j-(\Delta v)_j/2$ to $v_j+(\Delta v)_j/2$ will be referred to the jth level in the analogue to digital conversion, and therefore the mean square error voltage σ_j^2 associated with this level is

$$\sigma_j{}^2 = \int_{v_j - \frac{(\Delta v)_j}{2}}^{v_j + \frac{(\Delta v)_j}{2}} (v - v_j)^2 \, p(v) \, dv$$

If the step size is small compared with the total amplitude range, we may assume that the signal is uniformly distributed over the step regardless of its statistical distribution over the complete range. Therefore $p(v) = p(v_j)$ and is a constant in the above integral, hence

$$\sigma_j{}^2 = \frac{(\Delta v)_j{}^3 \, p(v_j)}{12}$$

The total mean square quantizing noise voltage σ^2 is the sum of all the mean square error voltages introduced at each level, i.e.

$$\sigma^2 = \sum_j \sigma_j{}^2 = \frac{1}{12} \sum_j (\Delta v)_j{}^3 \, p(v_j)$$

$$= \frac{1}{12} \sum_j (\Delta v)_j{}^2 [p(v_j) \, (\Delta v)_j] \qquad (7\text{-}1)$$

The probability of the signal amplitude being within the jth step is

$$P_j = \int_{v_j - \frac{(\Delta v)_j}{2}}^{v_j + \frac{(\Delta v)_j}{2}} p(v) \, dv$$

and since $(\Delta v)_j$ is small compared with the total amplitude range,

$$P_j \approx p(v_j) \, (\Delta v)_j$$

and therefore Eq. (7-1) becomes

$$\sigma^2 \approx \frac{1}{12} \sum_j (\Delta v)_j{}^2 P_j \qquad (7\text{-}2)$$

When all the steps are of equal size, i.e. *linear quantization*,

$$\sum_j (\Delta v)_j{}^2 P_j = (\Delta v)^2$$

Therefore

$$\sigma_{\text{lin}}{}^2 = \frac{(\Delta v)^2}{12} \qquad (7\text{-}3)$$

Example 7-1. To give an indication of the magnitude of the quantizing noise power in relation to a given signal power, consider a sine wave of peak-to-peak amplitude $2V$ quantized into M levels. For linear quantization the step size $\Delta v = 2V/M$ and therefore the r.m.s. signal-to-quantizing noise voltage ratio is

$$\frac{V}{\sqrt{2}} \div \frac{\Delta v}{\sqrt{12}} = \frac{V}{\sqrt{2}} \times \frac{M\sqrt{12}}{2V} = 1\cdot225M$$

$$[7\text{-}1]$$

This ratio is tabulated below for various values of *M*.

Table 7-1

Quantizing levels M	No. of digits in binary code	r.m.s. signal-to-quantizing noise ratio		Approx. % distortion
			dB	
7	3	8·6	18·7	12
15	4	18·4	25·3	5·4
31	5	38·0	31·6	2·6
63	6	77·2	37·7	1·3
127	7	155	43·8	0·6

When converting speech signals into digital form, a significant reduction in quantizing noise can be realised by employing non-uniform quantization to suit the amplitude distribution of the signal; there is a greater probability of small amplitudes than of large ones (see section 7-1.3), and the probability density function peaks about the average value of zero volts; therefore a tapered step system, as shown in Fig. 7-2, provides an improvement over uniform quantization employing the same number of levels. There is also the added advantage that weaker signals, which do not occupy the full amplitude range of $+V$ to $-V$, will be quantized by considerably more levels; the accommodation of loud and quiet talkers in the same system is dealt with later in the chapter.

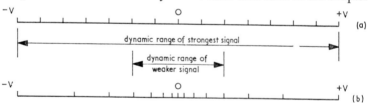

Fig. 7-2 (a) Uniform, and (b) non-uniform quantization of the signal amplitude range.

Before we consider the optimum tapering of the step size, which can only be determined from a knowledge of the amplitude distribution of the signal, let us refer again to Eq. (7-2), but this time assuming that the step associated with the jth level is one of a non-uniformly quantized arrangement. The mean square quantizing error voltage associated with this level can be written as

$$\sigma^2 \approx \frac{\overline{(\Delta v)^2}}{12} \tag{7-4}$$

where $\overline{(\Delta v)^2}$ is the mean square step size; for linear quantization $\overline{(\Delta v)^2} = (\Delta v)^2$, as shown previously by Eq. (7-3).

7–1.2 *Methods of Achieving Non-uniform Quantization*

There are, in the main, two methods of achieving non-uniform quantization. One is to compress the amplitudes of the analogue pulses in a non-linear amplifier followed by a linear encoder, i.e. an encoder that generates coded outputs for *equal* step inputs: the compression of the signal is known as *companding*. It must be stressed that for this application instantaneous compandors are required and they must not be confused with syllabic compandors used to protect weaker signals against

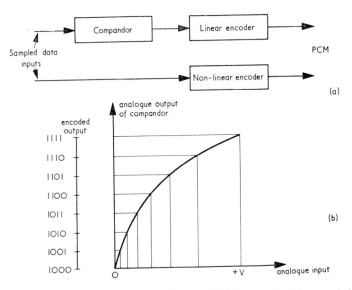

Fig. 7–3 (a) Non-uniform encoders for PCM, and (b) input-output characteristic of either the analogue compander or the non-linear encoder (shown for positive inputs only and with an encoded output representing half of a fifteen-level system).

noise over long-haul HF links.[1] To establish the theory of non-uniformly quantized speech signals we need not consider the design and circuitry of the compressor and encoder, but simply refer to the required input-output characteristic—a typical curve is shown in Fig. 7–3. Before describing this characteristic, it is desirable to introduce the second method of achieving non-uniform quantization; with this method the compression of the signal is carried out as an integral part of the encoding process and is known as non-linear encoding—further details can be

1. Theory of Syllabic Compandors, R. O. Carter, *Proc. I.E.E.*, Vol. 111, No. 3, pp. 503–13, March 1964.

[7–1]

found in the references.[2, 3] For our purposes the input-output characteristic can be described by the graph of Fig. 7–3, as for the first method of quantizing. Whichever method of quantization is used, a complementary expandor is required at the receiver to produce an overall linear compressor-expandor characteristic.

A compression curve that is reasonably flexible and relatively easy to implement is the logarithmic characteristic first suggested by Cattermole.[4] The normalised output y, i.e. maximum amplitude of unity, and normalised input x are related by the logarithmic expression

$$y = \frac{1 + \ln Ax}{1 + \ln A} \quad \text{for } 1/A \leqslant x \leqslant 1 \tag{7-5}$$

and a linear expression

$$y = \frac{Ax}{1 + \ln A} \quad \text{for } 0 \leqslant x \leqslant 1/A \tag{7-6}$$

where A, the compression coefficient, is a constant chosen to suit the amplitude distribution of the signal. The linear expression is required for low levels, i.e. for $Ax < 1$, to satisfy the condition that $y = 0$ when $x = 0$. Furthermore, in practice it is not possible to maintain a logarithmic characteristic at low levels which also justifies the use of the two expressions—note that they are continuous at $x = 1/A$, as illustrated in Fig. 7–4. It is customary to approximate the desired compression curve with a multilinear-segmental characteristic, which can be achieved in the analogue compression process of the first method of non-uniform quantization or in the non-linear encoding of the second method. The input and output step sizes Δx and Δy respectively are determined by the encoding process, which is designed according to the required number of quantizing levels.

Let us now determine the quantizing noise that would arise from using this combined linear/non-linear quantization characteristic. The number of quantizing levels is sufficiently large for us to write, using Eqs. (7–5) and (7–6),

for the linear law, $\qquad \Delta y = \dfrac{A \, \Delta x}{1 + \ln A} \tag{7-7}$

and for the logarithmic law $\quad \Delta y = \dfrac{1}{1 + \ln A} \dfrac{\Delta x}{x} \tag{7-8}$

2. PCM System for Junction Telephone Circuits, A. D. Stevens, *Point-to-Point Telecommunications*, Vol. 10, No. 2, pp. 6–25, February 1966, and Vol. 10, No. 3, pp. 40–64, June 1966.

3. Integrated Circuits and Pulse Coding, A. E. Chatelon, *Electronics*, Vol. 39, No. 19, pp. 139–48, September 19, 1966.

4. A contribution to discussion on three papers on PCM Transmission of Speech, K. W. Cattermole, *Proc. I.E.E.*, Vol. 109B, No. 48, p. 486, November 1962.

Also $$\varDelta y = 2/M \qquad (7\text{-}9)$$

since 2 is the normalised amplitude range and M the number of levels. If the analogue signal is sufficiently small to occupy the linear part of the characteristic only, then from Eqs. (7–3), (7–7) and (7–9) the mean square quantizing noise voltage is

$$\sigma_{\text{lin}}^2 = \frac{(\varDelta x)^2}{12} = \frac{(1+\ln A)^2}{3M^2 A^2} = \frac{k^2}{A^2} \qquad (7\text{-}10)$$

where the subscript 'lin' denotes linear quantization and

$$k = (1+\ln A)/M\sqrt{3}$$

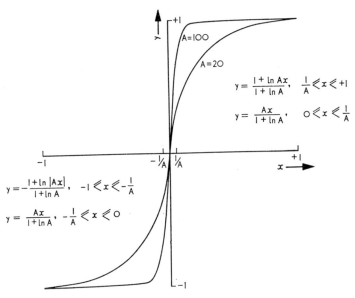

Fig. 7-4 Logarithmic encoding characteristic.

Next, if the analogue signal occupies the logarithmic portions of the characteristic only, a physically unrealizable condition but one which is convenient as an intermediate step in the analysis, then from Eqs. (7–4), (7–8) and (7–9) the mean square quantizing noise voltage is

$$\sigma_{\text{log}}^2 = \frac{\overline{(\varDelta x)^2}}{12} = \frac{(1+\ln A)^2}{3M^2}\,\overline{x^2} = k^2\overline{x^2} \qquad (7\text{-}11)$$

where $\overline{x^2}$ is the mean square amplitude of the signal and the subscript 'log' signifies logarithmic quantization.

[7–1]

A physically realizable signal will give rise to noise from both linear and non-linear quantization; therefore we wish to find

    (i) the proportion of the time for which the signal amplitude is within the range $-1/A < x < 1/A$, for a given r.m.s. value, and

    (ii) the effective mean square amplitude of the signal when occupying the amplitude ranges -1 to $-1/A$ and $+1/A$ to $+1$, i.e. $\overline{x^2}$ to account for the logarithmic quantizing noise.

We shall consider speech signals in the next section.

7–1.3 *Quantization of Speech Signals*

The variation of instantaneous level of a speech signal has been studied by Davenport,[5] who suggests that it can be represented as a uniform probability distribution for very low levels plus a negative exponential distribution for the high-level sounds. A reasonable approximation[6, 7] is possible using only the negative exponential term which is given by the probability density function

$$p(x) = \frac{1}{\sigma_s \sqrt{2}} \exp\left(-\frac{\sqrt{2}|x|}{\sigma_s}\right) \qquad (7\text{-}12)$$

where σ_s is the r.m.s. speech amplitude; note that $p(x)$ is symmetrical about $x = 0$.

The probability distribution $P(-1/A < x < 1/A)$, which is the first of the requirements stated in the last paragraph of section 7–1.2, is given by

$$P(-1/A < x < 1/A) = \int_{-1/A}^{1/A} p(x)\, dx$$

$$= 2\int_{0}^{1/A} \frac{1}{\sigma_s \sqrt{2}} \exp\left(-\frac{\sqrt{2}|x|}{\sigma_s}\right) dx$$

$$= 1 - \exp\left(-\frac{\sqrt{2}}{\sigma_s A}\right) \qquad (7\text{-}13)$$

Note that $1/A$ and hence x and σ_s, are normalised amplitudes—the maximum signal range being -1 to $+1$.

5. Experimental Study of Speech Wave Distributions, W. B. Davenport, *J. Acous. Soc. America*, 24, pp. 390–9, July 1952.

6. Instantaneous Companding of Quantized Signals, B. Smith, *Bell System Tech. J.*, 36, pp. 653–703, May 1957.

7. A Survey of Telephone Speech-Signal Statistics, R. F. Purton, *Proc. I.E.E.*, 109B, 43, pp. 60–6, January 1962.

The second requirement, namely to find $\overline{x_{\log}^2}$, is satisfied by

$$\overline{x_{\log}^2} = 2\int_{1/A}^{1} x^2\, p(x)\, dx$$

$$= 2\int_{1/A}^{1} \frac{x^2}{\sigma_s\sqrt{2}} \exp\left(-\frac{\sqrt{2}\,|x|}{\sigma_s}\right) dx$$

Integrating by parts and assuming $1/A \ll 1$, which will be shown to be a reasonable approximation, gives

$$\overline{x_{\log}^2} = \left(\frac{1}{A^2} + \frac{\sqrt{2}\sigma_s}{A} + \sigma_s^2\right) \exp\left(-\frac{\sqrt{2}}{\sigma_s A}\right) \qquad (7\text{–}14)$$

and therefore, using Eqs. (7–10) and (7–11), the total quantization noise is

$$\sigma_n^2 = \sigma_{\lin}^2 + \sigma_{\log}^2 = k^2\left[\frac{1}{A^2} P\left(-\frac{1}{A} < x < \frac{1}{A}\right) + \overline{x_{\log}^2}\right] \qquad (7\text{–}15)$$

The signal-to-quantizing noise power ratio is

$$\frac{S}{N_q} = \left[\frac{\sigma_s}{\sigma_n}\right]^2 = \frac{\sigma_s^2}{k^2\left[\dfrac{1}{A^2} P\left(-\dfrac{1}{A} < x < \dfrac{1}{A}\right) + \overline{x_{\log}^2}\right]}$$

or, for the purpose of illustration (see Fig. 7–5), the normalised signal-to-quantizing noise power ratio is

$$\frac{k^2\sigma_s^2}{\sigma_n^2} = \frac{\sigma_s^2}{\dfrac{1}{A^2} P\left(-\dfrac{1}{A} < x < \dfrac{1}{A}\right) + \overline{x_{\log}^2}} \qquad (7\text{–}16)$$

For example, when $\sigma_s = 1/A$,

$$\frac{k^2\sigma_s^2}{\sigma_n^2} = \frac{1/A^2}{\left(\dfrac{1}{A^2}(0\cdot76) + \dfrac{1}{A^2}(1+\sqrt{2}+1)(0\cdot24)\right)} = \frac{1}{1\cdot58} \equiv -2 \text{ dB}$$

We see from Fig. 7–5 that the signal-to-quantizing noise ratio is constant for r.m.s. signal amplitudes greater than $1/A$, ignoring the 2 dB droop at the lower levels, and therefore a system capable of handling a range of inputs should be designed for the r.m.s. amplitude of the quietest talker equal to $1/A$. From observations on commercial telephone circuits a signal-to-quantizing noise ratio of not less than 26 dB has been found acceptable and therefore k must be chosen to make the normalised ratio of 0 dB equivalent to 26 dB. Firstly, however, it is necessary to know the

[7–1]

likely range of mean amplitudes in order that peak clipping of the loudest talker may not be troublesome. It has been established by a number of sources, see Reference 7 for a summary and bibliography, that the mean speech amplitudes have approximately a Gaussian distribution with a standard deviation between 4 and 6 dB; when this

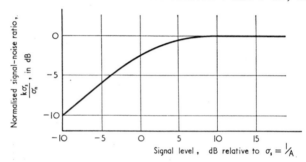

Fig. 7–5 Normalised signal-to-quantizing noise ratio against r.m.s. speech level.

figure is 5·5 dB, 98 % of the talkers have mean amplitudes within ± 13 dB of the median talker and 99·8 % are within ± 17 dB. Therefore the PCM system must handle a mean-level range of approximately 30 dB.

Fig. 7–6, which is based upon Eq. (7–12), indicates that for a given speech signal the probability of exceeding a power level of 13 dB above

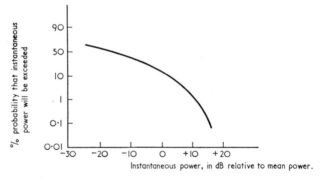

Fig. 7–6 Percentage probability that instantaneous power level of a speech signal will be exceeded.

the mean level is only 1 %; we shall use this as the criterion for determining the normalised r.m.s. amplitude of the loudest talker in order that peak clipping may be negligible. The *useful volume range* (UVR) must therefore extend from $1/A$ to 13 dB below the maximum normalised amplitude of unity, i.e.

$$\text{UVR} = [20(\log_{10}A) - 13] \text{ dB} \qquad (7\text{--}17)$$

This gives an indication of the required value of A, the compression coefficient, since the UVR is to be approximately 30 dB—hence A must be of the order of 150.

Having fixed the r.m.s. amplitudes in relation to the normalised range (see Fig. 7–7), it is now necessary to estimate the size of the smallest step, which is the step size Δx throughout the linearly quantized range, to satisfy the required 26 dB signal-to-quantizing noise ratio for the quietest talker, whose r.m.s. amplitude is $1/A$. Using Eq. (7–13), $P(x<|1/A|) \approx 0.75$ for $\sigma_s = 1/A$, indicating that the signal is uniformly quantized for three-quarters of the time. However, to simplify the analysis we shall assume the quietest talker to be linearly quantized all

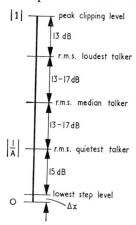

Fig. 7–7 r.m.s. amplitudes of speech signals in relation to the dynamic range of the non-uniform encoding system.

of the time, a somewhat loose approximation, but a justifiable one if we are concerned only with estimating the likely size of the smallest step. Let the r.m.s. signal amplitude be $r\Delta x$; using Eq. (7–3), the r.m.s. signal-to-quantizing noise ratio is $r\Delta x/(\Delta x/\sqrt{12})$, which must be equal to 20— i.e. the 26 dB requirement—therefore $r \approx 20/\sqrt{12} = 5.8$, which means that the r.m.s. amplitude of the quietest talker must be 15 dB greater than the lowest step level.

The total amplitude range between the lowest step level and the peak clipping level is therefore $15+30+13 = 58$ dB, which means that the smallest step is $1/800$ of 1 (1 being the normalised peak amplitude) and hence from Eqs. (7–7) and (7–9) we have

$$\Delta x = \frac{2(1+\ln A)}{MA} = \frac{1}{800}$$

$$[7\text{–}1]$$

or for a given compression coefficient A, the minimum number of levels to satisfy the requirement $\Delta x \leqslant 1/800$ is given by

$$M = \frac{1600\,(1+\ln A)}{A} \qquad (7\text{--}18)$$

Note that for $A = 1$, that is linear quantization throughout the whole range, the minimum number of levels would have to be 1600, which would require a 10-unit binary code. But for $A = 150$, M need only be 63 and hence a 6-unit binary code would suffice.

Finally, let us consider the choice of A and M to meet the 26 dB signal-to-quantizing noise ratio that is required for the whole range of signals. It can be seen from Eq. (7–16), which is illustrated in Fig. 7–5, that this ratio is approximately equal to a constant and given by

$$\frac{\sigma_\text{s}}{\sigma_\text{n}} \approx \frac{1}{k} \quad \text{for } 1/A < \sigma_\text{s} < 1$$

$$= \frac{M\sqrt{3}}{1+\ln A} \qquad (7\text{--}19)$$

The interdependence of M and A and their effect upon the signal-to-noise ratio is shown in Fig. 7–8; the usable volume range and its con-

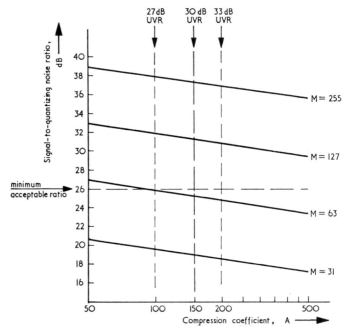

Fig. 7–8 Signal-to-quantizing noise ratio against compression coefficient for various numbers of quantizing levels.

nection with the compression coefficient, see Eq. (7–17), is also marked on the graph. We conclude that a 30 dB UVR and a minimum signal-to-quantizing noise ratio of 26 dB can only be achieved by a system employing 127 levels and a compression characteristic of not less than 150.

7–2 Delta Modulation (ΔM)

The essential difference between PCM and ΔM is that in the former information of the instantaneous amplitude of the signal is transmitted as an n-digit code, whereas in the basic ΔM system the rate of change of signal amplitude is communicated and in delta-sigma modulation $(\Delta-\Sigma M)$ the instantaneous amplitude is transmitted, but both systems employ single-digit codes.

7–2.1 *Delta Modulator employing 'Staircase' Integration*

As an introduction to the topic it is convenient to consider the ideal system first studied by de Jager[8] and shown in Fig. 7–9 (*a*). The transmitter consists of a comparator/modulator whose input is the difference

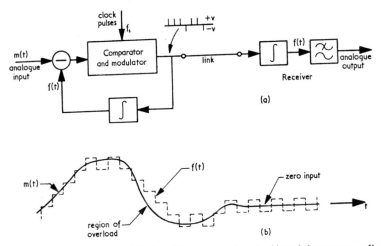

Fig. 7–9 (*a*) Basic delta modulation. (*b*) Analogue input $m(t)$ and the corresponding output $f(t)$ from the staircase integrator.

between its integrated output and the input signal $m(t)$. The modulator is triggered by clock pulses and, when the input signal amplitude $m(t)$ is greater than the integrated output, a positive output pulse is produced; conversely, when $m(t)$ is smaller than the feedback signal, a negative

8. Deltamodulation, a Method of PCM Transmission using a 1-unit Code, F. de Jager, *Philips Research Report*, Vol. 7, pp. 442–66, 1952.

[7–2]

output pulse is generated. It can be seen from Fig. 7–9 (*b*) that the integrated output is of a stepped form which follows the smooth continuous input $m(t)$. Therefore a similar integrator at the receiving terminal, followed by a low-pass filter to smooth out the abrupt changes of the stepped waveform, will produce an output resembling the original message $m(t)$; the original and reconstructed signals will not be identical and we have, as for PCM, quantization noise. Referring again to Fig. 7–9 (*b*) we see that when the input $m(t)$ is zero the digital output consists of alternate positive and negative pulses, which will have a discrete line spectrum with the first line at the pulse repetition frequency f_s. It is shown in the following analysis that f_s is much greater than the highest frequency in the input signal $m(t)$, and therefore, for this special condition of zero input, the output from the receiver low-pass filter will also be zero.

An important aspect of delta modulation is the problem of overloading which occurs whenever the input changes too rapidly for the stepped waveform to follow it; we see from Fig. 7–9 (*b*) that this occurs when the transition from one step to the next does not cross the input waveform $m(t)$. Let the amplitudes of the modulator output pulses be $+V$ and $-V$, and the height of each step be σ and if overloading is to be avoided

$$\frac{d[m(t)]}{dt} \times t_s \leqslant \sigma \qquad (7\text{--}20)$$

where $t_s = 1/f_s$, the time interval between pulses. For the purpose of illustration, a sine-wave input $m(t) = E_m \sin\omega_m t$ would require the condition

$$\omega_m E_m \frac{1}{f_s} \leqslant \sigma$$

i.e.
$$f_m E_m \leqslant \frac{\sigma f_s}{2\pi} \qquad (7\text{--}21)$$

if overloading is to be avoided. At this stage it is not possible to consider the implications of increasing V (i.e. increasing σ) and/or f_s in order that a higher frequency or larger amplitude of the input signal may be accommodated, but it would appear that increasing the pulse rate f_s should provide the best solution, since an increase in the pulse amplitude V would result in an integrated output made up of larger steps, thereby increasing the quantization noise accompanying the reconstructed signal —this is pursued later. It has been found experimentally[9] that a *Δ*M system can handle speech signals without overloading provided that the instantaneous amplitude of the signal does not exceed the peak ampli-

9. See Reference 8.

tude of a 800 Hz sine wave which can also be transmitted without over-loading taking place; this is because the energy of the speech signal is not uniformly distributed over the band 300–3400 Hz—the higher frequencies contain less energy than the lower ones.

7–2.2 *Delta Modulator employing RC Feedback and Full-width Pulses*

Let us now consider a practical system in which an RC network is used in place of a staircase integrator and full-width pulses replace the pseudo-impulses of the previous system. The exponential responses of the RC network to positive and negative step functions of amplitude $2V$ are shown as AB and CD in Fig. 7–10. The slopes of the curves vary over the range $+V$ to $-V$, but, since the time constant $T = RC$ is large compared with the pulse duration t_s, the output from the RC network, which follows the analogue input, may be considered as a piecewise linear waveform with the slope of each segment determined by the appropriate exponential curve and the amplitude of the waveform in relation to the dynamic range $+V$ to $-V$ (this is illustrated in the figure for both zero and sine-wave inputs). The slope of the exponential curve at amplitude v is $(V-v)/T$ and the slope of the sine wave at the same amplitude is $\omega_m \sqrt{E_m^2-v^2}$. The condition for which overloading is avoided is defined as

$$\frac{V-v}{T} \geqslant \omega_m \sqrt{E_m^2-v^2}$$

or
$$D = \frac{V-v}{T} - \omega_m \sqrt{E_m^2-v^2} \geqslant 0 \qquad (7\text{--}22)$$

i.e. the difference D of the slopes must always be positive and at the limit equal to zero. Differentiating D with respect to v, it is found to be a minimum when $v = E_m/\sqrt{1+\omega_m^2T^2}$; substituting this expression and the condition $D = 0$ into Eq. (7–22), the overload condition for a sine-wave input is set by

$$E_{max} = \frac{V}{\sqrt{1+\omega_m^2T^2}} \qquad (7\text{--}23)$$

where E_{max} signifies E_m for $D = 0$.

The amplitude and frequency of the sine-wave input illustrated in Fig. 7–10 satisfies the overload condition of Eq. (7–22) and corresponds to the limit $D = 0$; the integrated and input waveforms do not cross or touch each other at regular intervals, which was the required condition to avoid overloading in the system employing staircase integration (refer to Fig. 7–9 (*b*)).

[7–2]

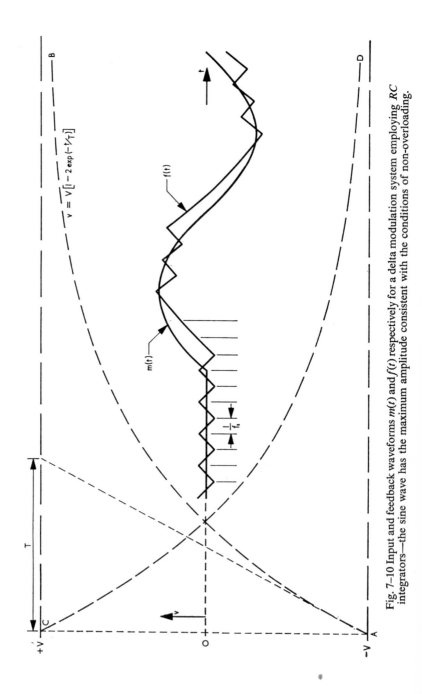

Fig. 7–10 Input and feedback waveforms $m(t)$ and $f(t)$ respectively for a delta modulation system employing RC integrators—the sine wave has the maximum amplitude consistent with the conditions of non-overloading.

$v = V\left[1 - 2\exp(-t/_T)\right]$

A plot of the overload characteristic of Eq. (7–23), which is identical to the steady-state amplitude-frequency characteristic of the *RC* network, is shown in Fig. 7–11 together with the amplitude-frequency spectrum of natural speech.[10] In practice, it is desirable to work as closely as possible to the overload condition in order that the signal-to-noise ratio may be kept as high as possible; we see from the relative positions of the two curves that this is achieved at the expense of some overloading of the middle frequency range.

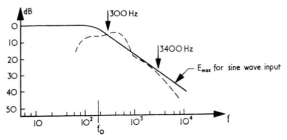

Fig. 7–11 Overload characteristic of a basic delta modulator employing single *RC* integration—broken curve represents the amplitude spectrum of natural speech.

When the peak amplitude of the sine-wave input is smaller than the central step size Δv the modulator idling pattern of positive and negative pulses is undisturbed and hence zero information is transmitted—this is referred to as the *threshold of coding* and is considered in detail in the following section.

7–2.3 *Quantization Noise*

Unlike PCM, the evaluation of the quantization noise in a delta-modulation system is extremely complex. Rigorous analyses are available,[11, 12] but the student meeting the subject for the first time is advised to consult these references after he is thoroughly conversant with the appendices of this book.

Quantization noise is defined as the difference between the filtered output signal and the input signal, which according to the definitions illustrated in Fig. 7–12 has a spectrum defined by

$$Q(j\omega) = M(j\omega) - G(j\omega)$$

10. Détermination du Spectre de la Parole avec une Méthode Nouvelle, T. Tarnóczy, *Acustica*, Vol. 8, pp. 392–5, 1958.

11. Quantization Noise of a Single Integration Delta Modulation System with an N-digit Code, H. van de Weg, *Philips Research Report*, Vol. 8, pp. 367–85, 1953.

12. Spectra of Quantized Signals, W. R. Bennett, *Bell System Tech. J.*, Vol. 27, pp. 446–72, July 1948.

Also $E(j\omega) = M(j\omega) - F(j\omega) = M(j\omega) - S(j\omega)H(j\omega)$

or $E(j\omega)A(j\omega) = M(j\omega) - S(j\omega)H(j\omega)A(j\omega)$

and assuming $A(j\omega)$ is flat over the spectrum of $M(j\omega)$,

$$E(j\omega)A(j\omega) = M(j\omega) - G(j\omega) \text{ and}$$

Therefore, $Q(j\omega) = E(j\omega)A(j\omega)$

The difficulty arises in determining $E(j\omega)$; even with a single sine-wave input the resulting pulse train $s(t)$ is random and consequently $E(j\omega)$ is the Fourier transform of a non-deterministic waveform. This method of analysis is not pursued further and has only been introduced to illustrate the complex nature of a rigorous treatment.

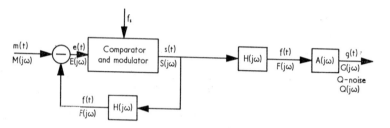

Fig. 7–12 Basic delta modulation.

Referring again to Fig. 7–10, the error waveform $e(t)$, i.e. the quantization noise before filtering, is triangular when the input $m(t)$ is zero. Before inspecting the error corresponding to the sine-wave input, it is necessary to re-draw the integrated and input waveforms with the latter delayed by a clock pulse interval t_s; this is evident from an inspection of the figure, also bearing in mind that the output can only respond to a change of the input after it has taken place—the re-drawn waveforms and their difference $e(t)$ are shown in Fig. 7–13. Although the error $e(t)$ is random for the sine-wave input, there is a distinct triangular pattern somewhat similar to that which occurs for zero input. It must be realised that in these illustrations the sine wave represents two extremes; firstly, it has the maximum permissible amplitude set by the conditions to avoid overloading and, secondly, it represents the highest input frequency that would be accommodated in a practical system —referring again to Fig. 7–10, the clock rate is twenty times the sine-wave frequency and it will be seen later that this is about the smallest allowable figure for the ratio $f_s : f_m$. Thus we may conclude that the triangularity of the error waveform will be more pronounced for all other acceptable inputs.

Let us consider the error waveform $e(t)$ resulting from zero input; the slope of each segment is V/T and the step height Δv, i.e. the peak-to-peak amplitude, is

$$\Delta v \approx \frac{V}{Tf_s} = \frac{2\pi V f_0}{f_s} \tag{7-24}$$

where $f_0 = 1/2\pi T$, the 3 dB break frequency of the RC network.

A similar expression can be obtained by a different line of reasoning. The digital output corresponding to zero analogue input is a sequence of alternate $+V$ and $-V$ pulses having a fundamental frequency $f_s/2$. The periodic sequence has a line spectrum, the first being at $f_s/2$, and since $f_s \gg f_0$ we may consider that the fundamental sine-wave component is responsible for exciting the RC network, thus producing the error waveform $e(t)$. Therefore the output response of the general network to this digital input is given by the approximation

$$f(t) \approx \frac{4V}{\pi} \sin\left(\frac{2\pi f_s t}{2}\right) \left| H\left(j\frac{\omega_s}{2}\right) \right|$$

and the slope $f'(t) \approx 4V f_s \cos(\pi f_s t) \left| H\left(j\frac{\omega_s}{2}\right) \right|$

Therefore the step height Δv, is

$$f'(t)\bigg|_{\max} \times \frac{1}{f_s} \approx 4V \left| H\left(j\frac{\omega_s}{2}\right) \right| \tag{7-25}$$

and for the RC network, assuming $f_0 \ll f_s$, we have

$$\Delta v \approx 4V\frac{2f_0}{f_s} = \frac{4}{\pi}\frac{2\pi V f_0}{f_s} \tag{7-26}$$

which is similar to the previous expression, Eq. (7-24), except for the coefficient $4/\pi$.

The mean square error is

$$\overline{e^2(t)} = \frac{2}{t_s}\int_0^{t_s/2}\left[\frac{(\Delta v)t}{t_s}\right]^2 dt = \frac{(\Delta v)^2}{12} \tag{7-27}$$

and we have seen previously that this power will be contained in a line spectrum at multiples of f_s. Next consider a sine-wave input; it is reasonable to assume that the mean square error voltage is proportional to $(\Delta v)^2$, since the triangular characteristic of the waveform is similar to that corresponding to zero input. It can also be seen from Fig. 7-13 that the error is greater for the sine wave, and it has been established[13] by

13. Calculating Delta Modulator Performance, F. B. Johnson, *Trans. I.E.E.E.*, Vol. AU-16, No. 1, pp. 121–9, March 1968.

[7-2]

computer analysis that the mean square error for various sine-wave inputs is $\overline{e^2(t)} \approx (\Delta v)^2/6$ and that this remains constant over a range of input amplitudes from just above the threshold of coding to the overloading level.* But this error power is associated with a random-sequence triangularly shaped polar function which, according to the analysis of section 8–4, page 194, has a continuous spectrum and a power density that is proportional to $|G(j\omega)|^2$. A triangular pulse of peak amplitude A and duration $2t_p$ has a Fourier transform

$$G(j\omega) = At_p \left[\frac{\sin(\omega t_p/2)}{\omega t_p/2} \right]^2$$

(see Tutorial Exercise I–3, page 222); the total energy of the pulse is

$$2A^2 t_p/3 \text{ volt}^2\text{-seconds}$$

and at $\omega = 0$, $|G(j\omega)|^2 = |G(0)|^2 = A^2 t_p^2$. Therefore we can say that the energy contained in the positive (one-sided) frequency spectrum is $A^2 t_p/3$ and is equal to the energy that would result from the spectral density at the origin, i.e. $A^2 t_p^2$, if it were held constant over a bandwidth from 0 to $+1/3t_p$. In the present application $1/t_p \equiv f_s$ and hence the unfiltered quantization noise energy may be considered to be uniformly distributed over a bandwidth from 0 to $f_s/3$. Therefore, if the receiver low-pass filter has a bandwidth from 0 to B_m, where $B_m \ll f_s$, the mean quantization noise power N_q is given by

$$N_q \approx \overline{e^2(t)} \frac{3B_m}{f_s} = \frac{(\Delta v)^2}{6} \frac{3B_m}{f_s}$$

and using Eq. (7–24), we have

$$N_q \approx \frac{2\pi^2 V^2 f_0^2 B_m}{f_s^3} \tag{7–28}$$

If we assume that the pulses supplied to the receiver decoder are of amplitude $+V$ and $-V$, a condition that can easily be satisfied by pulse regeneration, but which may cause some digits to be in error (see section 8–2 for further discussion), the mean square signal amplitude of the output corresponding to the limit of non-overloading at the input of the system is, according to Eq. (7–23), $E_{max}^2/2 = V^2/2(1+\omega_m^2 T^2)$; hence the mean signal-to-quantizing noise power ratio is

$$\frac{S}{N_q} \approx \frac{f_s^3}{4\pi^2(1+\omega_m^2 T^2)f_0^2 B_m} \tag{7–29}$$

---

* Reference 14, page 158, quotes $\overline{e^2(t)} \approx (\Delta v)^2/7$ for a single *RC* network in the feedback circuit.

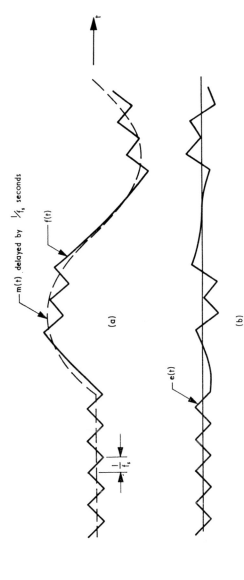

Fig. 7-13 (a) Delayed sine-wave input and integrated waveform. (b) Error waveform, $e(t) = m(t) - f(t)$.

Referring to Fig. 7–11, we see that f_0, the RC integrator characteristic frequency, is approximately equal to the lowest frequency of the speech signal and therefore in general $f_m \gg f_0$, and hence $\omega_m^2 T^2 \gg 1$, which allows Eq. (7–29) to be written as

$$\frac{S}{N_q} \approx \frac{f_s^3}{4\pi^2 f_m^2 B_m} \approx 0.025 \frac{f_s^3}{f_m^2 B_m} \qquad (7\text{–}30)$$

The form of this expression agrees with that of de Jager (see Reference 8), and also with more recent work,[14] except for the coefficient 0·025, which is 0·04 and 0·067 for the respective references. Bearing this in mind, together with the approximations that have been made in the above analysis, the results should not be interpreted too rigidly.

It is tempting to compare ΔM and PCM for a sine-wave input that, in the former, satisfies the limiting condition of non-overloading and, in the second, has a peak-to-peak amplitude occupying the full quantized amplitude range. In fact, this has been carried out on a number of occasions and usually a plot of signal-to-noise ratio against clock rate shows that ΔM is superior to PCM, when the latter uses a code group of five or less. Unfortunately such a comparison can be misleading if no account is taken of the range of input levels that must be handled by a viable system. It has been shown in section 7–1.3 that speech signals occupy a 30 dB mean level range from the quietest to loudest talkers and considerable attention has been given to the design of a PCM system capable of handling such a range. Using Eq. (7–28) we see that, for the ΔM system described, an output signal-to-quantizing noise ratio of 34 dB is obtained for a 800 Hz sine-wave input, which we shall assume from previous comments is representative of speech, an output bandwidth $B_m = 3$ kHz, and a clock rate of 56 kilobits/second—the latter is the PCM digit rate when using a 7-digit code, i.e. it is equal to 7 times the sampling rate of the analogue input, which is 8000 per second (see page 132). It has been stated previously that the signal-to-quantizing noise ratio for commercial telephony should not be less than 26 dB (see page 145), and therefore the useful dynamic range of the single RC integrator ΔM system is $34 - 26 = 8$ dB, which is very much smaller than the useful volume range attainable by PCM operating at the same overall digit rate (see Eq. (7–19) and Fig. 7–8).

Although this form of delta modulation is unsuitable for commercial telephony, there are possible applications in other fields such as telemetry for which the dynamic range need not be so large or the per-

14. Companded Delta Modulation for Telephone Transmission, A. Tomozawa and H. Kaneko, *Trans. I.E.E.E.*, Vol. COM-16, No. 1, pp. 149–57, February 1968.

missible signal-to-quantizing noise ratio may be less than 26 dB. With this in mind let us pursue the analysis further.

The maximum amplitude of the sine-wave input consistent with the condition of non-overloading has been established, see Eq. (7–23). The *minimum* peak amplitude below which the input fails to excite the modulator occurs when the peak-to-peak amplitude of the signal is smaller than the central step size Δv; this results in a modulator output of alternate positive and negative pulses corresponding to zero input. Let us define this lower amplitude limit (or *threshold of coding*) as $E_{min} = (\Delta v)/2$ where E_{min} is the peak amplitude of the sine-wave input. Using Eqs. (7–24) and (7–23), we have

$$E_{min} = \frac{\Delta v}{2} = \frac{\pi V f_0}{f_s}$$

and

$$E_{max} = \frac{V}{\sqrt{1+\omega_m^2 T^2}}$$

Therefore

$$E_{min} = \frac{\pi f_0}{f_s} E_{max} \sqrt{1+\omega_m^2 T^2}$$

and substituting

$$T = 1/2\pi f_0$$

then

$$\frac{E_{max}}{E_{min}} = \frac{f_s}{\pi \sqrt{f_0^2 + f_m^2}} \tag{7–31}$$

The ratio of E_{max}/E_{min}, which is usually expressed in dB, is the dynamic input range of the system; for the previous numerical example in which $f_s = 56$ kilobits/second, $f_m = 800$ Hz and $f_0 \approx 150$ Hz (see Fig. 7–11), $20 \log_{10}(E_{max}/E_{min}) = 28$ dB. The quantizing noise is independent of the input signal amplitude and therefore the signal-to-quantizing noise ratio varies from 34 to 7 dB over the dynamic range; whether or not a signal-to-noise ratio of 7 dB is acceptable would depend upon the particular application.

7–2.4 *Delta-sigma Modulation* $(\Delta\text{-}\Sigma M)$[15]

The system so far described exhibits an overload characteristic that is particularly suited to speech when using a high-quality microphone. It is not so attractive when using a carbon microphone or for video signals such as television. A system that provides an overloading characteristic that is independent of input frequency—particularly attractive for inputs

15. A Telemetering System by Code Modulation—Delta-sigma Modulation, H. Inose, Y. Yasuda and J. Murakami, *Trans. I.R.E.*, Vol. SET-8, pp. 204–9, September 1962.

[7–2]

with relatively flat energy spectra—is known as *delta-sigma* modulation. It is similar to the previous system, except that an integrator $H(j\omega)$ is included in the input circuit and a complementary inverse process $H^{-1}(j\omega)$ in the output circuit; however, it is shown that the latter may be combined with the receiver integrator to allow both to be dispensed with.

The inclusion of an integrator before what has so far been considered a delta modulator means that information of the amplitude of the input and not the derivative of the signal is transmitted. This raises the question of what is the difference between $\Delta\text{-}\Sigma M$ and PCM—simply, the latter involves an ADC process and the generation of a n-digit code, whereas $\Delta\text{-}\Sigma M$ is a single-digit code system and furthermore it does not require the complex encoding and decoding associated with PCM; this is discussed in more detail later in the chapter.

Let us pursue an analysis along similar lines to that of basic delta modulation. It is firstly necessary to determine the maximum amplitude of a sine-wave input consistent with the condition of non-overloading. The slope of the output from the first RC integrator is

$$\frac{dv}{dt} = \frac{\omega_m E_m \cos\omega_m t}{\sqrt{1+\omega_m^2 T^2}} = \frac{\omega_m}{\sqrt{1+\omega_m^2 T^2}} \sqrt{E_m^2 - v^2(1+\omega_m^2 T^2)}$$

and the difference D between this slope and the slope of the output from the feedback RC integrator is

$$D = \frac{V-v}{T} - \frac{\omega_m}{\sqrt{1+\omega_m^2 T^2}} \sqrt{E_m^2 - v^2(1+\omega_m^2 T^2)} \qquad (7\text{--}32)$$

which must be greater than or at the limit equal to zero to avoid overloading. The zero limit is found by differentiating D with respect to v and equating to zero, which gives $v = E_m/(1+\omega_m^2 T^2)$. Substituting this and $D = 0$ into Eq. (7–32) gives the non-overloading condition as

$$E_m = E_{\max} = V \qquad (7\text{--}33)$$

i.e. a maximum peak amplitude of the input equal to the pulse amplitude—a result that the reader may have realised from an inspection of the block diagram of Fig. 7–14 without recourse to the above procedure. Nevertheless, whichever method is adopted it can be seen that the overload condition is independent of the input sine-wave frequency f_m.

The quantization noise in a $\Delta\text{-}\Sigma M$ system can be assessed from the analysis of section 7–2.3. It has been shown that the error power density at the input to the receiver low-pass filter is approximately

$$\frac{(\Delta v)^2}{6} \frac{3}{f_s} \text{ watts/Hz}$$

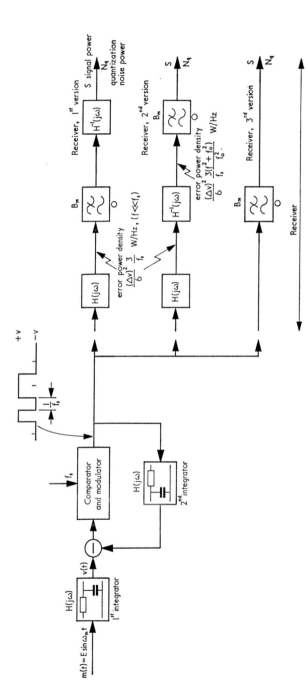

Fig. 7-14 Block diagram of the delta-sigma modulation system.

Rearranging the order of the receiver processes—which is permissible with linear networks—(see Fig. 7–14), second version of receiver decoder, the error power density at the output of the network $H^{-1}(j\omega)$ is

$$\frac{(\Delta v)^2}{6}\frac{3}{f_s}\left|H^{-1}(j\omega)\right|^2 = \frac{(\Delta v)^2}{6}\frac{3}{f_s}\left(\frac{f^2+f_0^2}{f_0^2}\right) \text{ watts/Hz}$$

where the symbols have the same meaning as in the previous section. The quantizing noise power N_q at the final output is therefore

$$N_q \approx \int_0^{B_m} \frac{(\Delta v)^2}{6}\frac{3}{f_s}\left(\frac{f^2+f_0^2}{f_0^2}\right) df$$

$$= \frac{(\Delta v)^2}{2f_s f_0^2}\left(\frac{B_m^3}{3}+B_m f_0^2\right)$$

Using Eq. (7–24) and substituting for Δv,

$$N_q \approx \frac{2\pi^2 V^2}{f_s^3}\left(\frac{B_m^3}{3}+B_m f_0^2\right) \tag{7–34}$$

We see that the quantization noise is reduced by making f_0 as small as possible, i.e. the time constant T of the RC network as large as possible. If we satisfy this requirement, $B_m \gg f_0$ and

$$N_q \approx \frac{2\pi^2 V^2}{3}\left(\frac{B_m}{f_s}\right)^3, \quad f_0 \ll B_m \tag{7–35}$$

Next let us evaluate the mean signal power at the output. Referring to Fig. 7–14 and the first version of the receiver decoder, the signal power at the output of the low-pass filter is approximately equal to the signal power at the comparator input, namely $\overline{v^2(t)} = E_m^2/[2(1+\omega_m^2 T^2)]$—this line of reasoning was used in the previous section dealing with the basic ΔM system. Therefore the signal power at the output of the Δ-ΣM receiver is

$$\frac{E_m^2}{2(1+\omega_m^2 T^2)}\left|H^{-1}(j\omega_m)\right|^2 = \frac{E_m^2}{2(1+\omega_m^2 T^2)}(1+\omega_m^2 T^2) = \frac{E_m^2}{2}$$

which is simply the mean power at the input of the system. Hence, the mean output power corresponding to the limit of non-overloading at the input is $E^2_{max}/2 = V^2/2$ (see Eq. (7–33)). Thus the mean signal-to-quantizing noise ratio is

$$\frac{S}{N_q} \approx \frac{3}{4\pi^2}\left(\frac{f_s}{B_m}\right)^3, \quad f_0 \ll B_m \tag{7–36}$$

As an illustration consider a sine-wave input to a system having a bandwidth from d.c. to B_m of 3 kHz—strictly speaking, this is not a system to handle speech, as this would employ a band-pass filter in the receiver in place of a low-pass unit and consequently the lower limit of the integral leading up to Eq. (7–34) would be 300 and not zero. The signal-to-quantizing noise ratio corresponding to $f_s = 56$ kilobits/second and $B_m = 3$ kHz, is 28 dB; we see that this is only marginally above the required ratio of 26 dB for commercial telephony. According to the analysis leading up to Eq. (7–34), the quantization noise is independent of the signal level, and since the 27 dB ratio corresponds to the maximum input it may be concluded that there is virtually zero dynamic range available for this application.

Let us pursue the general analysis to establish the dynamic range from the maximum input consistent with non-overloading to the minimum input corresponding to the threshold of coding. For the latter, let the input $m(t) = E_{min}\sin\omega_m t$; the peak output from the first RC network is $E_{min}/\sqrt{1+\omega_m^2 T^2}$, and therefore using the definition of the threshold of coding stated in the previous section dealing with basic ΔM, we obtain

$$\frac{E_{min}}{\sqrt{1+\omega_m^2 T^2}} = \frac{\Delta v}{2} = \frac{\pi V f_0}{f_s}$$

Combining this with Eq. (7–33), we have

$$\frac{E_{max}}{E_{min}} = \frac{f_s}{\pi \sqrt{f_0^2 + f_m^2}} \tag{7–37}$$

which is identical to Eq. (7–31) for basic ΔM.

The main advantage of Δ-ΣM is the ability to convey information of d.c. levels. For zero input there are an equal number of positive and negative pulses in the transmitted sequence, and hence the filtered output at the receiver is zero. For a positive d.c. input there are more positive pulses than negative ones, the average number being in direct proportion to the magnitude of the input level; similarly, there are more negative than positive pulses in the digital signal when the d.c. input is negative. This is true for both ΔM and Δ-ΣM, but with the latter the receiver is simply a low-pass filter which allows the average amplitude of the digital signal to be retained in the final output.

The choice of the optimum value of f_0, the 3 dB cut-off frequency of the RC network, highlights another difference between the two methods of modulation. In basic ΔM, f_0 must be chosen in accordance with the spectrum of the input signal as shown in Fig. 7–11, whilst with Δ-ΣM Eqs. (7–34) and (7–37) indicate that f_0 should be made as small as possible; in practice, the lower limit is set by the instabilities of the comparator—the lower f_0 the smaller the step size.

[7–2]

7–2.5 *Companded Delta Modulation*

The main drawback to basic ΔM and Δ-ΣM is the limited dynamic range over which a signal-to-quantizing noise ratio in excess of 26 dB is obtainable. It has been seen that this is due to the quantization noise being independent of the signal level and set solely by the quantizing step size Δv. The following system known as *companded delta modulation* employs a local decoder to modify the step size in accordance with the level of the input signal. There are basically two methods of achieving this. One, is to obtain a control signal whose magnitude is proportional to the mean level of the analogue input and to use this signal to amplitude modulate the digital sequence fed to the integrator in the feedback circuit[16]—it has been shown in section 7–2.3 that the step size is directly proportional to the amplitude of the pulses. The second method is similar except that the control signal is derived entirely from the digital output of the encoder; a brief account of a system[17] falling into the latter category is given below.

A block diagram of the scheme is shown in Fig. 7–15. The feedback circuit consists of an integrator $H_1(j\omega)$ which performs a similar function to those described previously, except that the input pulses are not of constant amplitude $\pm V$, but have a variable level $\pm C$ which is determined by a control signal amplitude modulating the digital output of the encoder. The control voltage is made up of two components, a d.c. bias which is included to prevent C falling below a level consistent with stable operation—the smaller C the smaller the step size and the higher must be the gain in the comparator—and the other, the output from a level detector which follows the integrator $H_2(j\omega)$. The time constant of the level detector is made sufficiently long for the control signal to respond only to syllabic changes. The relevant equations can be obtained from the analysis of basic ΔM. Let the analogue input be a sine wave $m(t) = E \sin\omega_m t$ and E_{max} represent its peak amplitude corresponding to the limit of non-over-loading. Using Eq. (7–23), but expressing the integrator transfer function in its general form, we have $E_{max} = V|H_1(j\omega_m)|$. In companded delta modulation the overload condition is defined similarly except that the amplitude of the pulses fed to the integrator is C; hence we have

$$E_{max} = C|H_1(j\omega_m)| \qquad (7\text{--}38)$$

When the input exceeds the overload limit the digital output is a continuous sequence of either $+V$ or $-V$ pulses. Let the ratio of the signal

16. Companded Delta Modulation for Telephony, S. J. Brotin and J. M. Brown, *Trans. I.E.E.E.*, Vol. COM-16, No. 1, pp. 157–62, February 1968.
17. See Reference 14.

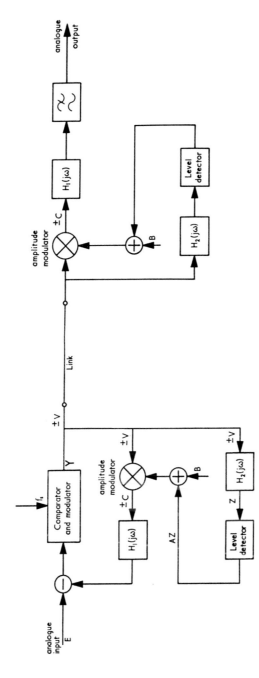

Fig. 7-15 Block diagram of a companded delta modulation system.

peak amplitude E to the overload level at the same frequency be expressed as

$$\mathsf{E} = \frac{E}{E_{\max}}, \quad E < E_{\max} \qquad (7\text{--}39)$$

Similarly, let the ratio of the signal level contained in the digital output to that corresponding to the overload level be

$$\mathsf{Y} = \left[\frac{E}{E_{\max}}\right]^n \qquad (7\text{--}40)$$

The reason for introducing the index n will become apparent as we proceed with the analysis—briefly, this is a non-linear encoder and the ratio of signal levels contained in the digital output is not linearly related to the ratio of the analogue levels. Note that when $E = E_{\max}$, $\mathsf{Y} = 1$. The output from the integrator $H_1(j\omega)$, which must follow the input level E, can be written as

$$E = C|H_1(j\omega_m)|\mathsf{Y} \qquad (7\text{--}41)$$

The output from the integrator $H_2(j\omega)$ is

$$Z = V|H_2(j\omega_m)|\mathsf{Y} \qquad (7\text{--}42)$$

and also

$$C = B + AZ \qquad (7\text{--}43)$$

where A is a coefficient of the level detector. Combining Eqs. (7–41), (7–42) and (7–43) in order that Z and C may be eliminated, we get

$$E = \{B + AV|H_2(j\omega_m)|\mathsf{Y}\}|H_1(j\omega_m)|\mathsf{Y} \qquad (7\text{--}44)$$

which means that Y is a function of the square root of the input level E, indicating that the encoder has a square-root-type compression characteristic; alternatively, it may be said that the local decoder has a square-type expanding characteristic.

For ease of manipulation it is convenient to normalise all of the variables in the above equations, as we have already done for $E/E_{\max} = \mathsf{E}$. Eq. (7–41) becomes

$$\frac{E}{E_{\max}} = \frac{C|H_1(j\omega_m)|}{V|H_1(j\omega_m)|}\mathsf{Y}$$

or

$$\mathsf{E} = \mathsf{C}\,\mathsf{Y} \qquad (7\text{--}45)$$

where $C/V = \mathsf{C}$, V being the maximum value of C. Similarly, the normalised version of Eq. (7–42) is

$$\frac{Z}{Z_{\max}} = \frac{V|H_2(j\omega_m)|}{V|H_2(j\omega_m)|}\mathsf{Y}$$

i.e.

$$\mathsf{Z} = \mathsf{Y} \qquad (7\text{--}46)$$

and Eq. (7–43) may be written as

$$\frac{C}{V} = \frac{B}{V} + \frac{AZ}{V} \tag{7-47}$$

and when C is a maximum,

$$1 = \frac{B}{V} + \frac{AV|H_2(j\omega_m)|}{V}$$

therefore $A = (1 - B/V)/|H_2(j\omega_m)|$

Substituting this result in Eq. (7–47) and putting $B/V = \mathsf{B}$, gives

$$\mathsf{C} = \mathsf{B} + \frac{(1-\mathsf{B})}{|H_2(j\omega_m)|} \frac{Z|H_2(j\omega_m)|}{Z_{max}}$$

i.e.
$$\mathsf{C} = \mathsf{B} + (1-\mathsf{B})\mathsf{Y} \tag{7-48}$$

Combining Eqs. (7–45) and (7–48) we obtain

$$\mathsf{E} = \mathsf{B}\mathsf{Y} + (1-\mathsf{B})\mathsf{Y}^2 \tag{7-49}$$

The effect of varying B is illustrated in Fig. 7–16—the smaller B the smaller the gradient of the curve and the greater the dynamic range.

The signal-to-quantizing noise power ratio can be evaluated from Eq. (7–28), but with V replaced by C, the control pulse amplitude, i.e.

$$N_q \approx \frac{2\pi^2 C^2 f_0^2 B_m}{f_s^3}$$

Using Eq. (7–41) to substitute for C and putting

$$|H(j\omega_m)| = f_0^2/(f_0^2 + f_m^2)$$

$$N_q \approx \frac{2\pi^2 E^2 B_m(f_0^2 + f_m^2)}{f_s^3 \mathsf{Y}^2} \tag{7-50}$$

Hence the mean signal-to-quantizing noise power ratio is

$$\frac{S}{N_q} \approx \frac{E^2}{2} \frac{f_s^3 \mathsf{Y}^2}{2\pi^2 E^2 B_m(f_0^2 + f_m^2)} \tag{7-51}$$

and for $\mathsf{Y} = 1$, i.e. for maximum amplitude input, it is identical to the basic ΔM relationship of Eq. (7–29). The difference between companded and basic delta modulation is the increased dynamic range of the former; with numerical values of $B_m = 3$ kHz, $f_0 = 150$ Hz, $f_m = 800$ Hz and $f_s = 56$ kilobits/second, the maximum signal-to-quantizing noise ratio is 34 dB. When the ratio is 26 dB, then from Eq. (7–51) $\mathsf{Y} = 0.4$ and

[7–2]

assuming $B = 0\cdot1$ Eq. (7–49) gives $E = 0\cdot184$, i.e. a 15 dB dynamic range as compared with 8 dB for basic ΔM (see page 158).

A higher maximum signal-to-quantizing noise ratio and a larger dynamic range can be achieved by using a double integrator as the feedback network $H_1(j\omega)$. A suitable arrangement is shown in Fig. 7–17, and for a system handling speech, typical network characteristics are $f_1 \approx 150$ Hz—similar to the choice made in the basic single integrator system—$f_2 \approx 1$ kHz and $f_3 \approx 9$ kHz (see Reference 14). We may assume that $f_1 \ll f \ll f_3$ and for $f_s = 56$ kilobits/second, $f_3 = f_s/2\pi$; consequently the network transfer function can be written as

$$|H(j\omega)| \approx \frac{f_1}{f\sqrt{1+(f/f_2)^2}} \tag{7-52}$$

and

$$\left|H\left(j\frac{\omega_s}{2}\right)\right| \approx 4\sqrt{1+\pi^2}\frac{f_1 f_2}{f_s^2} \tag{7-53}$$

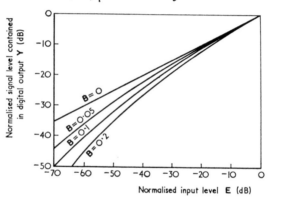

Fig. 7–16 Compression characteristic for various values of **B**.

The step response of this network is slower than that of the single *RC* integrator and hence the quantizing step size Δv is smaller. Using Eq. (7–25), but replacing V by C in order that it may represent a companded system, we have

$$\Delta v \approx 4C\left|H\left(j\frac{\omega_s}{2}\right)\right|$$

which may be combined with Eq. (7–53) to give

$$\Delta v \approx 16C\sqrt{1+\pi^2}\frac{f_1 f_2}{f_s^2} \tag{7-54}$$

Whereas the error signal $e(t)$ is related to the step size by the relationship $\overline{e^2(t)} \approx (\Delta v)^2/6$ for single integration (see page 156), it has been found that for double integration using the network of Fig. 7–17, the relationship is

$$\overline{e^2(t)} \approx \frac{(\Delta v)^2}{2 \cdot 2} \qquad (7\text{–}55)$$

Assuming that the quantization noise power is given by

$$N_q \approx \overline{e^2(t)}\, \frac{3B_m}{f_s}$$

(see page 156), then in the companded ΔM system

$$N_q \approx \left\{ \frac{16C\sqrt{1+\pi^2}f_1 f_2}{f_s^2} \right\}^2 \frac{3B_m}{2 \cdot 2 f_s}$$

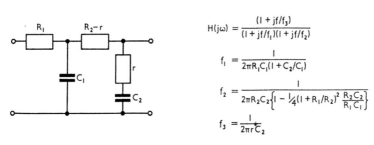

$$H(j\omega) = \frac{(1+jf/f_3)}{(1+jf/f_1)(1+jf/f_2)}$$

$$f_1 = \frac{1}{2\pi R_1 C_1(1+C_2/C_1)}$$

$$f_2 = \frac{1}{2\pi R_2 C_2\left\{1 - \frac{1}{4}(1+R_1/R_2)^2 \frac{R_2 C_2}{R_1 C_1}\right\}}$$

$$f_3 = \frac{1}{2\pi r C_2}$$

Fig. 7–17 Circuit diagram and transfer function of the double integration network.

Using Eq. (7–41) to substitute for C and Eq. (7–52) to replace $|H(j\omega_m)|$, the signal-to-quantizing noise power ratio is

$$\frac{S}{N_q} \approx \frac{E^2}{2} \frac{2 \cdot 64 \times 10^{-4} f_s^5 f_1^2 Y}{B_m f_1^2 f_2^2 E^2 f_m^2 [1+(f_m/f_2)^2]}$$

$$= 1 \cdot 32 \times 10^{-4} \frac{f_s^5 Y}{B_m f_m^2 (f_2^2 + f_m^2)} \qquad (7\text{–}56)$$

It is left to the reader to verify that the expression also applies to a non-companded double integration system when $Y = 1$. Using numerical values of $B_m = 3$ kHz, $f_m = 800$ Hz, $f_2 = 1$ kHz and $f_s = 56$ kilobits/second the signal-to-quantizing noise ratio corresponding to maximum input, i.e. $Y = 1$, is 43·5 dB. A 26 dB ratio corresponds to $Y = 0 \cdot 13$; substituting this value into Eq. (7–49) and assuming $B = 0 \cdot 1$ the analogue input ratio E is equal to 0·0286, or a dynamic range of $20 \log_{10}(1/0 \cdot 0286)$ = 31 dB.

[7–2]

Table 7–2

System $B_m = 3$ kHz, $f_s = 56$ kilobits/s	Maximum signal-to-quantizing noise ratio, dB	Dynamic range, in dB, for minimum signal-to-quantizing noise ratio of 26 dB
Single integration, basic ΔM	34	8
Double integration, basic ΔM	44	18
Single integration, $\Delta-\Sigma M$	28	2
Single integration, companded ΔM	34	15
Double integration, companded ΔM	44	31
Seven-digit companded PCM	30 dB UVR at a signal-to-noise ratio of 31 dB	

The characteristics of the systems described in this chapter are outlined in Table 7–2. It may be concluded that companded ΔM offers a specification comparable to PCM for the transmission of speech. It also offers the advantage of not requiring sophisticated encoders and decoders as in PCM and, furthermore, the same decoding network is used at both ends of the link, thereby obviating a matching problem, which is also a troublesome feature of a PCM system.

Tutorial Exercises

7–1 A 7-bit PCM system employing uniform quantization has an overall output digit rate of 100 kilobits/second.

Calculate the signal-to-quantizing noise ratio that would result when the input is a 1 kHz sine wave with peak-to-peak amplitude equal to the available amplitude range of the system.

Calculate the dynamic range for sine-wave inputs in order that the signal-to-quantizing noise ratio may not be less than 30 dB.

What is the theoretical maximum sine-wave frequency that the system can handle?

Ans: 43·8 dB; 13·8 dB; 7·14 kHz.

7–2 What are the advantages and disadvantages of transmitting a speech signal in digital form?

Suggest a suitable PCM method of quantizing the amplitude range of speech signals with particular reference to the quietest and loudest talkers.

The suggested method is used to quantize a sine wave whose peak-to-peak amplitude is equal to the maximum amplitude range of the system. Calculate he signal-to-quantizing noise power ratio.

Assume that the probability density function for the amplitude of the sine wave is

$$p(x) = \frac{1}{\pi\sqrt{2-x^2}}$$

where x is in units of the r.m.s. value.

Ans: $\dfrac{S}{N_q} \approx \dfrac{3M^2}{(1+\log A)^2}$

where M is the number of quantizing
levels, and
         A is the compression characteristic.

7-3 Show that a basic delta modulation system employing double integration (see Figs. 7-9 and 7-17) yields an output signal-to-quantizing noise power ratio

$$\frac{S}{N_q} \approx 1{\cdot}32 \times 10^{-4} \frac{f_s{}^5}{B_m f_m{}^2 (f_2{}^2 + f_m{}^2)}$$

when the input is a sine wave with maximum amplitude consistent with the conditions of non-overloading.

7-4 With the aid of courses on electronic circuit design, including digital and logical techniques, devise a system to sample speech and linearly encode the data to form a PCM signal. Carry out a review of the literature in order to establish the difference between designing linear and non-linear encoders.

Undertake a similar design study for a ΔM system. Compare the two systems in the light of present-day technology.

7-5 The output from a delta modulator employing a single RC integrator (time constant 1 millisecond) in the feedback circuit, is made up of $\pm3{\cdot}6$ volt NRZ pulses. The output rate is 60 kilobits/second.

Calculate the *maximum* peak amplitude of a 800 Hz sine wave in order that overloading does not occur. Also calculate the *minimum* peak amplitude of the sine wave corresponding to the threshold of coding. Hence deduce the dynamic range of the modulator.

Ans:—0·7 V, 30 mV, 27 dB.

8
DIGITAL SIGNALLING—BASEBAND ANALYSIS

It was indicated in Chapter 6 that digitally encoded information can tolerate more signal distortion due to intersymbol interference and noise pick-up than an analogue signal, and in Chapter 7 we have seen how the analogue-to-digital conversion can be effected. This chapter is concerned with establishing the optimum pulse shape and filtering of the digital signal, in order that the received pulse sequence may be interpreted with the minimum error. The overall transmission system is shown in Fig. 8–1 and the following analysis will deal with the baseband signal, i.e. a signal suitable for line communication; however, the basic principles apply equally to radio systems which require, in addition, modulators and demodulators operating at the radio carrier frequency.

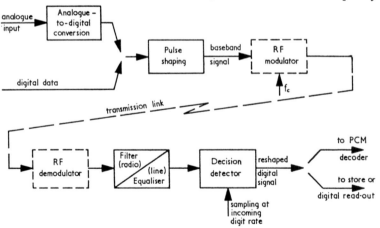

Fig. 8–1 Digital transmission system employing decision detection.

The essential difference between analogue and digital transmission is best illustrated by considering the function of the receiver. The relative amplitudes of the pulses of a PAM analogue transmission directly determine the characteristics of the reconstituted continuous signal, whereas, in a digital system, it is the coding (or grouping) of the digits that determines the analogue characteristics—the amplitudes of the pulses contributing indirectly. That is to say, it is only necessary for the receiver to decide the particular amplitude of a pulse out of a finite number of possibilities; once the decision has been made which, for a binary signal, is accomplished by deciding whether the amplitude of the signal-plus-noise is above or below a certain threshold level, the interference is virtually removed. There is, however, the possibility of introducing an error in the code whenever the interference masks the signal to such an extent that an incorrect decision is made. A method of minimising the error is to precede the *decision detector* with a filter (known as a *matched filter*), which maximises the signal-to-noise power ratio, and to arrange for the sampling of each pulse to occur at the instant when the signal power is at a maximum. There is also the requirement of zero crosstalk at the sampling instants and ideally the overall transfer function of the system should be chosen accordingly (see section 8–1 and refer to page 134). In the design of radio systems handling digital signals both of these requirements need to be considered, whereas in line systems the signal-to-noise ratio is sufficiently high to dispense with the matched filter; instead, the effective bandwidth of the line is widened by equalisation in order that a higher digit rate may be achieved.

The effect of making a wrong decision depends upon the kind of information being transmitted; an error of 1 in 10^3 could seriously impair computer data, whereas a similar error rate, associated with a PCM commercial telephony signal, would cause little or no disturbance to the restored analogue message. On circuits handling both PCM and general data the unprotected error rate, i.e. the digit-by-digit error liability due to interference and not the protected error rate of a system that includes error detection and correction, must be low, say 1 in 10^5 or better, and hence the disturbance to the reconstituted analogue information of the PCM signal is negligible. On circuits handling PCM only, the capacity may be increased—and hence the required bandwidth —to such an extent that the error rate becomes relatively high, say 1 in 10^2, and the resulting distortion of the analogue output can be significant; it is shown in section 8–3 that PCM suffers a threshold effect similar to other modulation systems.

Usually, for the transmission of computer data, an error rate of 1 in 10^5 is unacceptably high and the information-carrying digits must be protected with additional digits arranged in a coded sequence. Before deciding on a coding format it must be known whether the errors are likely to occur singly or in clusters. This is considered for a non-fading signal in the presence of band-limited noise of Gaussian distribution

[8]

and, although a somewhat idealised condition, it serves as a useful introduction to the *autocorrelation function* (see section III–2, page 271).

Apart from determining the optimum pulse shape to minimise crosstalk, there is also the question of deciding upon the polarity of the pulses for line systems. In a binary system, for example, the two states could be represented as pulse or no-pulse, i.e. an on-off signal, or possibly by equal amplitude pulses but of opposite polarity, known as a *polar* signal. It is shown in section 8–2 that for a given transmitter power the pulse polarity representation has a significant effect upon the digit error rate. It will also be seen that there is a marked difference between the frequency spectra of the signals; a random sequence of on-off pulses can have both a continuous and a discrete line spectrum, the latter being particularly useful for synchronising the decision detection sampling of the incoming digits. On the other hand, a polar pulse train only exhibits a continuous spectrum and the timing information must be extracted by first converting the pulse sequence into an on-off form.

8–1 Pulse Shape and Intersymbol Interference (Crosstalk)

It was shown on page 128 that crosstalk can be eliminated by using $\sin x/x$ pulses transmitted at a rate of twice the cut-off frequency of the low-pass filter that forms the pulses. The main drawback to the implementation of this method of signalling is the extreme accuracy to which the transmitting pulse rate and receiver sampling rate would have to be kept if the overlapping oscillatory tails are to cause no concern—there is also the practical difficulty of making an ideal low-pass filter! An alternative arrangement is to use a filter with a more gradual cut-off, thereby reducing the oscillatory tails of the output response whilst maintaining the overlapping zero crosstalk feature of the $\sin x/x$ response. A physically realizable filter meeting these requirements can be defined by an expression incorporating the ideal low-pass filter characteristic modified by a sinusoidal amplitude-frequency characteristic having odd-symmetry about the cut-off frequency (see Fig. 8–2). The amplitude-frequency response of the resultant transfer function may be defined as[1]

$$|H(j\omega)| = \tfrac{1}{2}\left[1+\sin\frac{\pi(\omega_0-\omega)}{2\omega_x}\right], \quad \omega_0-\omega_x<\omega<\omega_0+\omega_x$$

$$= 1, \qquad\qquad\qquad \omega<\omega_x$$

$$= 0, \qquad\qquad\qquad \omega>\omega_0+\omega_x$$

and for simplification of the mathematics a linear phase-frequency response is assumed, a condition which, in practice, can be approached more closely with this type of filter than it can for one with extremely

1. See Reference 8 (Chapter 6), page 135.

sharp cut-off. The amplitudes of the oscillatory tails of the output response are smallest when $\omega_x = \omega_0$, i.e. the most gradual cut-off (see

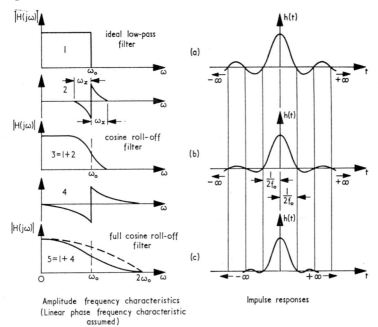

Amplitude frequency characteristics
(Linear phase frequency characteristic assumed)

Impulse responses

Fig. 8–2 Ideal low-pass and cosine roll-off filter characteristics and their impulse responses. (The dashed curve of (c) shows the filter characteristic having the impulse response $h(t)$ when the input is a full-width rectangular pulse.)

Fig. 8–2 (c)): this is known as the *full-cosine roll-off* characteristic and the amplitude-frequency response of the transfer function is

$$|H(j\omega)| = \left.\begin{array}{l} \dfrac{1}{2}\left[1+\cos\dfrac{\pi\omega}{2\omega_0}\right] \\[2mm] = \cos^2\dfrac{\pi f}{4f_0} \end{array}\right\} \quad 0<\omega<2\omega_0 \qquad (8\text{--}1)$$

The output response to a pseudo-impulse function with Fourier transform $F(j\omega) = k$, is

$$h(t) = \int_{-2f_0}^{2f_0} \frac{k}{2}\left[1+\cos\left(\frac{\pi\omega}{2\omega_0}\right)\right]\exp(j\omega t)\,df$$

$$= \frac{k\omega_0}{\pi}\frac{\sin 2\omega_0 t}{2\omega_0 t}\frac{1}{1-(2\omega_0 t/\pi)^2} \qquad (8\text{--}2)$$

$$[8\text{--}1]$$

and in Fig. 8–2 it can be seen how this response compares with the corresponding low-pass filter response, namely

$$h(t) = \frac{k\omega_0}{\pi} \frac{\sin\omega_0 t}{\omega_0 t} \qquad (8\text{–}3)$$

(see page 232). For zero crosstalk, the sampling rate $f_s = 2f_0$.

So far we have considered the filter whose *impulse* response is the required pulse shape for zero crosstalk. In practice, the pulse would not be generated from an impulse, since the amplitude of the latter would have to be exceedingly large for the output to contain a significant amount of energy. If, instead, rectangular pulses of, say, width t_p and amplitude V are used, the Fourier transform of the input is

$$F(j\omega) = Vt_p \frac{\sin\omega t_p/2}{\omega t_p/2}$$

and the amplitude-frequency characteristic of the required shaping filter is

$$|H(j\omega)| = \frac{|G(j\omega)|}{|F(j\omega)|}$$

where $G(j\omega)$ is the Fourier transform of the output response given by Eq. (8–2), i.e. $G(j\omega)$ must have an amplitude-frequency characteristic that is identical to the expression of Eq. (8–1); hence

$$|H(j\omega)| = \frac{\frac{1}{2}[1+\cos(\pi\omega/2\omega_0)]}{\sin(\omega t_p/2)/(\omega t_p/2)}, \quad 0 < \omega < 2\omega_0 \qquad (8\text{–}4)$$

(Note that Vt_p and k, both of which have the dimensions of volt-seconds, have been omitted.) As an example, for full-width rectangular pulses, $t_p = 1/f_s = 1/2f_0$ and the transfer function of the shaping filter is shown by the dashed curve in Fig. 8–2 (*c*).

It has been explained in the introduction to this chapter that noise can be removed from the received signal by sampling each digit when it has reached its maximum value, which is usually at the centre of the time slot for the digit. A predetection filter must not only maximise the signal-to-noise ratio but also produce an output that will satisfy the zero crosstalk requirements described above. Take, for example, a sequence of uncurbed rectangular pulses which is received with little or no pulse distortion, i.e. transmitted over a relatively wideband link or equalised line, but accompanied by noise of uniform spectral density. The zero crosstalk requirement could be satisfied by using a filter having the amplitude-frequency response defined by Eq. (8–4) and shown as the dashed curve in Fig. 8–2 (*c*). But would this filter maximise the peak signal-to-noise power ratio for the decision detector to produce a re-formed signal containing the minimum number of errors? The answer is no, but, to substantiate this comment, let us consider the general transmission system in which pulse shaping (or filtering) is included at both ends of the link (see Fig. 8–3). The input $f(t)$ corresponds to a

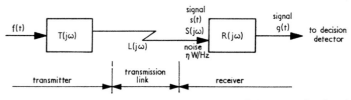

Fig. 8-3 Basic transmission system defined in terms of three transfer functions: $T(j\omega)$, $L(j\omega)$ and $R(j\omega)$.

single pulse of the digital signal; the associated output response $g(t)$ must satisfy the condition of zero crosstalk. The Fourier transform of $g(t)$ can be expressed as

$$G(j\omega) = F(j\omega)\, T(j\omega)\, L(j\omega)\, R(j\omega) \qquad (8\text{--}5)$$

where the various transfer functions are defined in Fig. 8–3. Since $g(t)$ is specified, $G(j\omega)$ is known and therefore

if $F(j\omega) = k$, i.e. a pseudo-impulse input,

$$T(j\omega)\, L(j\omega)\, R(j\omega) \equiv \frac{1}{2}\left[1+\cos\left(\frac{\pi\omega}{2\omega_0}\right)\right] \qquad (8\text{--}6)$$

and if $F(j\omega) = Vt_\mathrm{p}\sin(\omega t_\mathrm{p}/2)/(\omega t_\mathrm{p}/2)$, i.e. a rectangular pulse input

$$T(j\omega)\, L(j\omega)\, R(j\omega) \equiv \frac{\frac{1}{2}[1+\cos(\pi\omega/2\omega_0)]}{\sin(\omega t_\mathrm{p}/2)/(\omega t_\mathrm{p}/2)} \qquad (8\text{--}7)$$

We now wish to know if a proportion of the overall amplitude-frequency characteristic of the transmission system can be allocated to $R(j\omega)$ to maximise the signal-to-noise power ratio at the input to the decision detector. Consider the received signal, not including noise, to have an amplitude-frequency spectrum $S(j\omega)$—corresponding to a time-response $s(t)$ which at this stage need not be explicitly known. We have

$$G(j\omega) = S(j\omega)\, R(j\omega) \qquad (8\text{--}8)$$

and
$$g(t) = \int_{-\infty}^{\infty} S(j\omega)\, R(j\omega)\, \exp(j\omega t)\, df \qquad (8\text{--}9)$$

$$[8\text{--}1]$$

The one-sided noise power spectral density of the filter output is

$$\eta |R(j\omega)|^2 \text{ watts/Hz}$$

and the total output noise $N = \int_0^\infty \eta |R(j\omega)|^2 \, df$

$$\text{or} \qquad \int_{-\infty}^\infty \frac{\eta}{2} |R(j\omega)|^2 \, df.$$

We require to maximise the ratio $|g(t)|^2 / N$ at some instant in time, say $t = T_d$, i.e.

$$\frac{|g(T_d)|^2}{N} = \frac{\left| \int_{-\infty}^\infty S(j\omega) \, R(j\omega) \exp(j\omega T_d) \, df \right|^2}{\dfrac{\eta}{2} \int_{-\infty}^\infty |R(j\omega)|^2 df} \qquad (8\text{--}10)$$

The maximisation of this expression can be accomplished by applying the Schwarz inequality[2]

$$\int P^* \, P \, dx \int Q^* \, Q \, dx \geqslant |\int P^* \, Q \, dx|^2$$

where P^* is the complex conjugate of P; the equality applies when $P = constant \times Q$. For the present application let

$$P^* = S(j\omega) \exp(j\omega T_d)$$

$$\text{and } Q = R(j\omega)$$

But $S(j\omega) = S^*(-j\omega)$ and $R(j\omega) = R^*(-j\omega)$ for the input and output time responses to be real, therefore

$$\frac{|g(T_d)|^2}{N} \leqslant \frac{\displaystyle\int_{-\infty}^\infty |R(j\omega)|^2 df \int_{-\infty}^\infty |S(j\omega)|^2 df}{\dfrac{\eta}{2} \displaystyle\int_{-\infty}^\infty |R(j\omega)|^2 df}$$

i.e.

$$\frac{|g(T_d)|^2}{N} \leqslant \frac{2 \displaystyle\int_{-\infty}^\infty |S(j\omega)|^2 df}{\eta}$$

But $\int_{-\infty}^\infty |S(j\omega)|^2 df$ is equal to the energy contained in the pulse (see page 241); hence

$$\frac{|g(T_d)|^2}{N} \leqslant \frac{2 \, (Energy \ in \ pulse)}{one\text{-}sided \ noise \ power \ density} \qquad (8\text{--}11)$$

2. *Ancillary Mathematics*, H. S. W. Massey and H. Kestelman, p. 362 (Pitman, 1959).

and the equality applies, i.e. the signal-to-noise power ratio is a maximum at $t = T_d$, when

$$S^*(j\omega) \exp(-j\omega T_d) = constant \times R(j\omega) \qquad (8\text{-}12)$$

and the filter is said to be matched to the input signal—hence the term *matched filter*. Note that the constant in Eq. (8–12) has the dimensions of volt-seconds.

As a side issue, consider the impulse response of the matched filter, which is

$$h(t) = \int_{-\infty}^{\infty} R(j\omega) \exp(j\omega t) \, df$$

$$= \int_{-\infty}^{\infty} S^*(j\omega) \exp[j\omega(t-T_d)] \, df$$

When $S(j\omega)$ is real, i.e. the input $s(t)$ is symmetrical, $h(t) = s(t-T_d)$, which means that the impulse response is a delayed replica of the input signal $s(t)$ to which the filter is matched; see also Tutorial Exercise 8–1, page 202.

Using Eqs. (8–8) and 8–12),

$$G(j\omega) = \frac{1}{constant} |S(j\omega)|^2 \exp(-i\omega T_d) \quad \text{Vs} \qquad (8\text{-}13)$$

or

$$= constant |R(j\omega)|^2 \exp(-j\omega T_d) \quad \text{Vs} \qquad (8\text{-}14)$$

Again, note the dimensions of the constant term and that the equations are dimensionally correct. The amplitude-frequency characteristic of the pulse $g(t)$ having a 1 V peak amplitude and giving zero crosstalk is

$$|G(j\omega)| = \frac{1}{2}\left[1+\cos\left(\frac{\pi\omega}{2\omega_0}\right)\right] = \cos^2\left(\frac{\pi f}{4f_0}\right) \quad \text{volt-seconds}$$

and therefore from Eq. (8–14) the matched filter must have an amplitude-frequency response

$$|R(j\omega)| \equiv \cos\left(\frac{\pi f}{4f_0}\right) \qquad (8\text{-}15)$$

and, of course, a linear phase-frequency characteristic giving rise to a constant time delay T_d. Having established the characteristics of the receiving filter, it is now possible to determine $T(j\omega) \, L(j\omega)$ by using either Eq. (8–6) or Eq. (8–7), depending upon whether the input $f(t)$ is an impulse or rectangular pulse. The separation of the two transfer functions in order that $T(j\omega)$ may be determined requires a knowledge of the line characteristics $L(j\omega)$; even the assumption of a simple RC

[8–1]

model, as used in Chapter 6, would result in a complicated expression for $T(j\omega)$.

We are now beginning to see that the ideal condition of zero crosstalk, which is perfectly feasible when considered in terms of the output response of a network defined by a single transfer function, is of dubious implementation for a communication system made up of many stages in cascade. Nevertheless, the problem of maximising the signal-to-noise power ratio before decision detection and at the same time minimising the crosstalk is at the forefront of the design of any digital communication system.

For line systems the problem of designing the complicated filters postulated for zero crosstalk can be alleviated by using non-ideal pulses, e.g. rectangular, setting the digit rate according to the available bandwidth of the equalised line and allowing a finite but acceptable level of intersymbol interference or crosstalk. Regenerative repeaters, which provide the equalisation of the line characteristics, also maintain the signal-to-noise ratio at a relatively high level and hence a matched filter is not an essential requirement. To establish these points, consider the following example.

Example 8–1. A digital signal consisting of a sequence of uncurbed rectangular pulses is transmitted over a line that can be represented by the simple *RC* network previously introduced on page 134. The received distorted pulse (st) has a Fourier transform

$$S(j\omega) = F(j\omega)\,L(j\omega) = Vt_\text{p}\,\frac{\sin(\omega t_\text{p}/2)}{(\omega t_\text{p}/2)}\,\frac{1}{1+jf/f_\text{3dB}} \qquad (8\text{–}16)$$

The transfer function of the matched filter for this input $s(t)$ can be found from Eq. (8–12), but before attempting its evaluation let us examine the impulse response which, for an unsymmetrical input as $s(t)$ is in this example, is a delayed replica of $s(t)$ run backwards in time. Such an impulse response in unrealistic, since it commences at $t = -\infty$; therefore the matched filter cannot be realised and we are thus relieved of the task of determining its transfer function. Instead, assume that a sin x/x filter is used, which is the matched filter for an uncurbed rectangular input when $x = \omega t_\text{p}/2$ (see Exercise 8–2, page 202). The Fourier transform $G(j\omega)$ of the output $g(t)$ corresponding to the input $s(t)$ is

$$G(j\omega) = S(j\omega)\,R(j\omega)$$

$$= Vt_\text{p}\,\frac{\sin(\omega t_\text{p}/2)}{(\omega t_\text{p}/2)}\,\frac{1}{(1+jf/f_\text{3dB})}\,\frac{\sin(\omega t_\text{p}/2)}{(\omega t_\text{p}/2)}$$

An expression for $g(t)$ must be established in order that the peak power and crosstalk may be determined; the inverse Fourier integral is not recommended as a method of solution, but rather the use of the Convolution Integral,

$$g(t) = \int_{-\infty}^{\infty} s(\tau)h(t-\tau)d\tau$$

(see page 243). Referring to Fig. 8–4 and Exercise I–7, page 248, the received distorted pulse is defined as

$$s(\tau) \begin{cases} = V[1-\exp(-\tau/RC)], & 0<\tau<t_\text{p} & (8\text{–}17) \\ = V\exp(-\tau/RC)[\exp(t_\text{p}/RC)-1], & \tau>t_\text{p} & (8\text{–}18) \end{cases}$$

where V is the amplitude of the uncurbed rectangular pulse $f(t)$ and for the RC network simulating the line, $f_{3\text{dB}} = 1/2\pi RC$. The transfer function of the receiver filter is

$$R(j\omega) = \frac{\sin(\omega t_\text{p}/2)}{(\omega t_\text{p}/2)}\exp(-j\omega T_\text{d})$$

and its impulse response is

$$h(t) = \int_{-\infty}^{\infty} 1 \times \frac{\sin(\omega t_\text{p}/2)}{(\omega t_\text{p}/2)}\exp[j\omega(t-T_\text{d})]df$$

$$= \frac{1}{t_\text{p}}, \quad (T_\text{d}-t_\text{p}/2)<t<(T_\text{d}+t_\text{p}/2)$$

(Note that the t_p term in the denominator of the expression for $h(t)$ appears as a result of the definition $R(j\omega)$; $h(t)$ has the dimensions of volts, since $F_\delta(j\omega)$ = 1 volt-second. In the following analysis the filter time delay T_d has been ignored; however, it may be included simply by writing $(t-T_\text{d})$ in place of t.

There are four conditions of t for which the Convolution Integral must be evaluated (see Fig. 8–4 (b)):

(i) when $t<0$, $g(t) = 0$

(ii) when $0<t<t_\text{p}$,

$$g(t) = \int_0^t V[1-\exp(-\tau/RC)]\frac{1}{t_\text{p}}d\tau$$

$$= \frac{V}{t_\text{p}}\{t+RC[\exp(-t/RC)-1]\} \qquad (8\text{–}19)$$

and is a maximum when $t = t_\text{p}$.

(iii) when $t_\text{p}<t<2t_\text{p}$,

$$g(t) = \int_{t-t_\text{p}}^{t_\text{p}} V[1-\exp(-\tau/RC)]\frac{1}{t_\text{p}}d\tau$$

$$+ \int_{t_\text{p}}^{t} V\exp(-\tau/RC)[\exp(t_\text{p}/RC)-1]\frac{1}{t_\text{p}}d\tau$$

$$= \frac{V}{t_\text{p}}\{2t_\text{p}-t+RC[\exp(-t/RC)-2\exp[-(t-t_\text{p})/RC]+1]\} \quad (8\text{–}20)$$

Check that when $t = t_\text{p}$ this expression agrees with that for (ii).

(iv) when $t>2t_\text{p}$,

$$g(t) = \int_{t-t_\text{p}}^{t} V\exp(-\tau/RC)[\exp(t_\text{p}/RC)-1]\frac{1}{t_\text{p}}d\tau$$

$$= \frac{V}{t_\text{p}}\{RC[\exp(t_\text{p}/RC)-1][\exp[-(t-t_\text{p})/RC]-\exp(-t/RC)]\}$$

$$(8\text{–}21)$$

$$[8\text{–}1]$$

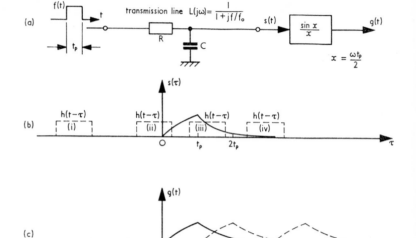

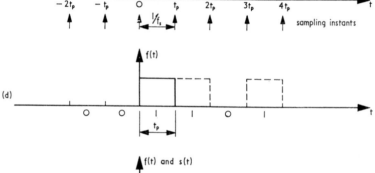

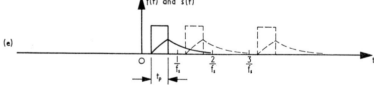

Fig. 8-4

(a) The transmission system.
(b) Illustration of the impulse response $h(t-\tau)$ of the sin x/x filter in relation to $s(\tau)$ for the evaluation of $g(t)$ using the Convolution Integral.
(c) Output response of the sin x/x filter, and
(d) the corresponding input $f(t)$ consisting of full-width pulses.
(e) Waveforms for half-width (50% duty cycle) pulses.

Check that when $t = 2t_p$ this expression agrees with that for (iii). The output $g(t)$ is a maximum when $t = t_p$, i.e.

$$g(t)_{max} = \frac{V}{t_p}\{t_p + RC[\exp(-t_p/RC)-1]\} \qquad (8\text{-}22)$$

Let us raise the question, albeit hypothetical, of how this filter compares with the matched filter, had it been physically realizable, in terms of the output signal-to-noise power ratio. To answer this question it is necessary to know the spectral power density of the noise at the input to the filters, which, for simplification of the analysis, can be assumed uniform and equal to η watts/Hz—a condition that is not valid for a practical system, but one which is justifiable for the present comparison.

The mean noise power output from the sin x/x filter is

$$N = \int_0^\infty \eta \left| \frac{\sin(\omega t_p/2)}{(\omega t_p/2)} \right|^2 df$$

and putting $\omega t_p/2 = x$,

$$N = \frac{\eta}{\pi t_p} \int_0^\infty \frac{\sin^2 x}{x^2}\, dx = \frac{\eta}{\pi t_p}\frac{\pi}{2} = \frac{\eta}{2t_p} \qquad (8\text{-}23)$$

The energy contained in the input pulse $s(t)$ is

$$E = \int_0^{t_p} V^2[1-\exp(-t/RC)]^2 dt + \int_{t_p}^\infty V^2\exp(-2t/RC)[\exp(t_p/RC)-1]^2 dt$$

$$= V^2\{t_p + RC[\exp(-t_p/RC)-1]\} \text{ volt}^2\text{-seconds} \qquad (8\text{-}24)$$

Using Eqs. (8-11), (8-22), (8-23) and (8-24) to compare the output signal-to-noise ratios of the matched and sin x/x filters, we have

$$\frac{2E}{\eta} \div \frac{[g(t)_{max}]^2}{N} = \frac{1}{1+RC/t_p[\exp(-t_p/RC)-1]} \qquad (8\text{-}25)$$

For a numerical comparison we must attach a figure to the ratio RC/t_p; assuming that full-width pulses are used (see Fig. 8-4 (d)), and the 3 dB cut-off frequency of the line is equal to half the sampling rate f_s, $t_p = 1/f_s$, $1/2\pi CR = 1/2t_p$ and hence $RC/t_p = 1/\pi$. Substituting this figure in the above equation yields a 1·6 dB degradation of the signal-to-noise ratio for the sin x/x filter compared with the match filter. Whether or not this is a significant reduction depends upon the actual signal-to-noise ratio; its effect in terms of the probability of making a wrong decision regarding the state of the signal at the specified instant in time is studied in the next section.

The second important issue is to determine the intersymbol interference (or crosstalk) which for full-width pulses is defined as the ratio of the maximum amplitude-to-amplitude of the pulse t_p seconds later (see Fig. 8-4 (c)). Using Eq. (8-21) and substituting $t = 2t_p$ (note the origin $t = 0$ coincides with the commencement of the pulse) and $RC/t_p = 1/\pi$, we get $g(2t_p) = 0.291$ V and using Eq. (8-22) the maximum pulse amplitude $g(t)_{max} = g(t_p) = 0.696$ V; hence the crosstalk ratio is 7·6 dB. Repeating the calculations for half-width

[8-1]

pulses, but again with $f_{3dB} = f_s/2$, $RC/t_p = 2/\pi$, the signal-to-noise ratio reduction when comparing the sin x/x and matched filter outputs is 3 dB and the crosstalk ratio, calculated as $g(t)_{max} : g(3t_p)$, is 15·6 dB—note, the smaller the intersymbol interference the higher the crosstalk ratio.

It is instructive to evaluate the crosstalk ratios of the pulses before filtering, i.e. for $s(t)$ in place of $g(t)$. Using Eq. (8–18), for the full-width pulse and $RC/t_p = 1/\pi$, $s(2t_p) = 0·042$ V and $s(t_p) = 0·95$ V giving a crosstalk ratio of 27 dB—cf. 7·6 dB corresponding to the filtered pulse $g(t)$. For half-width pulses and $RC/t_p = 2/\pi$, the crosstalk ratio before filtering is 27 dB, calculated as $s(t_p) : s(3t_p)$, compared with the 15·6 dB figure obtained for the filtered output. We conclude that filtering to maximise signal-to-noise ratio increases the intersymbol interference. *On line systems the signal-to-noise ratio can be kept relatively high and therefore the matched filter, or its nearest equivalent, can be dispensed with.*

Before leaving this section, let us examine the assumed relationship between the 3 dB cut-off frequency of the RC model of the line and the digit rate, namely $f_{3dB} = f_s/2$, which is analogous to the relationship between sampling rate and Nyquist bandwidth introduced in Chapter 6. Referring to the insertion loss-frequency characteristic for 20 SWG (5·6 kg/km) polyethylene insulated wires (Fig. 1–9, page 16), it can be seen that the 3 dB cut-off frequency for a 1·83 km (6000 feet) length is in the region of 30 kHz, which, according to the above condition, would impose a maximum limit on the digit rate of 60 kilobits/second. However, it is common practice in 24-channel PCM systems carrying speech information to use digit rates of the order of 1·5 Mbits/second over this type of cable by using equalisers in the repeaters that effectively extend the 3 dB (for a 1·83 km line) cut-off frequency from 30 kHz to the order of 300 kHz.[3] We conclude that in line systems it is not only essential to dispense with matched filters on the grounds of excessive crosstalk, but it is also desirable to extend the bandwidth of the line by equalisation—both conditions are dependent upon a high signal-to-noise power ratio being maintained throughout the link. These conclusions do not apply to radio systems in which the signal is more vulnerable to interference during transmission; filtering at the receiver to improve the signal-to-noise ratio is essential. This is considered to be outside the undergraduate curriculum and therefore beyond the scope of this book.

8–2 Decision Detection—Error Probability

The function of the decision detector is to determine the state of each digit out of a finite number of possibilities—two for a binary signal—in order that a re-formed noise-free pulse sequence may be generated; in a regenerative repeater the signal would be processed in this way for

3. Bipolar Repeater for PCM Signals, J. S. Mayo, *Bell Syst. Tech. J.*, Vol. 41, pp. 25–97, January 1962.

further transmission, whereas at the final destination the re-shaped digital signal would be either decoded, if it were a PCM or ΔM signal, or routed to the appropriate data-handling machine or store if general data. A typical on-off signal before transmission and the corresponding waveform at the input to the detector are shown in Fig. 8–5. The amplitude levels of the received signal if unaccompanied by noise would be

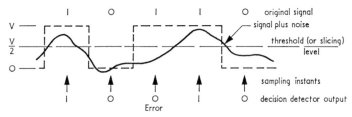

Fig. 8–5 Decision detection—the effect of noise on an on-off pulse transmission.

0 and V; assuming an equal probability of the two binary states, i.e. $p(0) = p(1) = \frac{1}{2}$, a *slicing* (or *threshold*) level is set at $V/2$. At the instant of inspection, which is made to occur when a pulse has reached its maximum amplitude, the detector decides whether the signal plus noise is above or below the slicing level and generates the appropriate output of 1 or 0. We are concerned with the noise masking the signal to such an extent that a wrong decision is made—as illustrated in the figure by the third digit being in error.

There are two conditions to be studied; the reception of no pulse, i.e. noise alone, or a pulse plus noise.

(i) *No pulse.* Noise alone will produce an error whenever its amplitude at the instant of inspection exceeds the slicing level. If we assume the noise to have a Gaussian distribution, the probability of an error is given by the area under the probability density curve (see Fig. 8–6 (*a*)), and is

$$P\left(v > \frac{V}{2}\right) = \int_{\frac{V}{2}}^{\infty} \frac{1}{\sqrt{2\pi N}} \exp\left[-\frac{v^2}{2N}\right] dv$$

where N is the mean noise power.

(ii) *Pulse plus noise.* An error will occur whenever the resultant amplitude is less than $V/2$ at the instant of inspection; the probability of an error is given by

$$P\left(v < \frac{V}{2}\right) = \int_{-\infty}^{\frac{V}{2}} \frac{1}{\sqrt{2\pi N}} \exp\left[-\frac{(v-V)^2}{2N}\right] dv$$

and is illustrated in Fig. 8–6 (*b*).

[8–2]

It can be seen from the symmetry of the curves that

$$P\left(v>\frac{V}{2}\right)\bigg|_{\substack{\text{noise}\\\text{alone}}} = P\left(v<\frac{V}{2}\right)\bigg|_{\substack{\text{pulse}\\\text{plus}\\\text{noise}}}$$

and since there is equal probability of 1 or 0 being transmitted, $P(v>V/2)$ can be used to determine the overall error probability P_e. Error function (erf) tables give numerical values of $P(-x_1<x<x_1)$ where x is in units of the r.m.s. value, or we may write

$$P(x>x_1) = \tfrac{1}{2}\{1-P(-x_1<x<x_1)\}$$
$$= \tfrac{1}{2}\{1-\text{erf }x_1\}$$

(see Appendix II, page 260). In the present application,

$$P_e = \frac{1}{2}\left\{1-P\left(-\frac{V}{2}<v<\frac{V}{2}\right)\right\}$$

where $V/2$ is the slicing level set at half the peak pulse amplitude. Before erf tables can be used it is necessary to convert v and V into units of the r.m.s. noise amplitude. We have

$$P\left(-\frac{V}{2}<v<\frac{V}{2}\right) = 2\int_0^{\frac{V}{2}} \frac{1}{\sqrt{2\pi N}} \exp\left(-\frac{v^2}{2N}\right) dv$$

and there are two well-used substitutions for the variable v.

1	2
let $\dfrac{v}{\sqrt{2N}} = y$	let $\dfrac{v}{\sqrt{N}} = y$
$2\displaystyle\int_0^{\frac{V}{2}} p(v)\,dv$	$2\displaystyle\int_0^{\frac{V}{2}} p(v)\,dv$
$= 2\displaystyle\int_0^{\frac{V}{2\sqrt{2N}}} \dfrac{1}{\sqrt{\pi}} \exp(-y^2)\,dy$	$= 2\displaystyle\int_0^{\frac{V}{2\sqrt{N}}} \dfrac{1}{\sqrt{2\pi}} \exp\left(-\dfrac{y^2}{2}\right)\,dy$
which is a tabulated function as	which is a tabulated function as
$\text{erf }x = \dfrac{2}{\sqrt{\pi}}\displaystyle\int_0^x \exp(-y^2)\,dy$	$\text{erf }x = \sqrt{\dfrac{2}{\pi}}\displaystyle\int_0^x \exp\left(-\dfrac{y^2}{2}\right)\,dy$
where $x = V/2\sqrt{2N}$	where $x = V/2\sqrt{N}$

$$P_e = \tfrac{1}{2}\{1-\text{erf }x\} \qquad (8\text{-}26)$$

Erf tables of both forms are available and care must be exercised to ensure which set is being consulted; a short form table using substitution 1 is to be found on pages 286–288. As a numerical illustration, when the peak signal-to-mean noise power ratio of an on-off signal is 17·4 dB, $V/\sqrt{N} = 7·45$, $x = 7·45/2\sqrt{2}$ and from the tables erf $x = 0·9998$ resulting in an error probability $P_e = 0·0001$ (1 in 10^4). Before plotting P_e against the signal-to-noise ratio, let us look at another form of signalling in which the two binary states are represented as equal positive and negative amplitude pulses, usually referred to as a *polar* signal. If a 1 is represented as $+V/2$ and a 0 as $-V/2$, the amplitude difference V is the same as for

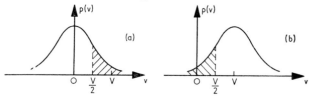

Fig. 8–6 Probability densities in on-off pulse transmission—shaded areas indicate probability of an error for (*a*) no pulse, and (*b*) pulse plus noise.

the on-off signal and, with equal probability of 1 and 0 occurring, the slicing level is set at zero volts. Under these conditions the probability of an error can be determined from the above analysis. A difference between the two forms of signalling and the common use of the analysis becomes apparent if we attempt to equate P_e to the *mean* signal-to-noise power ratio. For the on-off signal the peak pulse power is proportional to V^2 and the mean signal power proportional to $kV^2/2$ where k accounts for the pulse shape and the factor 1/2 arises, since the number of pulses is half the number of digits transmitted, assuming $p(1) = \frac{1}{2}$. In the polar system the peak pulse power is proportional to $(V/2)^2$ and, if we assume the same pulse shape as for the on-off signal, the mean signal power is proportional to $kV^2/4$, since there is a pulse for every digit transmitted. We conclude that for equal probability of an error the mean signal power of the polar signal is 3 dB less than that of the on-off signal.

It is often desirable to compare the performance of two or more systems operating under entirely different conditions of available signal power, digit rate and pulse shape; the signal-to-noise power ratios and error probabilities may be known either from measurement or assessed by calculation. The mean signal-to-noise power ratio can be expressed as

$$\frac{S}{N} = \frac{E f_s}{\eta B_n} \tag{8-27}$$

where E is the energy per pulse,

f_s is the digit rate,

B_n is the equivalent noise bandwidth (see page 91), and

η is the one-sided uniform spectral power density of the noise.

[8–2]

Re-arranging gives, by definition, the *normalised* signal-to-noise power ratio

$$R = \frac{E}{\eta} = \frac{average\ energy\ per\ pulse}{one\text{-}sided\ noise\ power\ density} = \frac{S}{N}\frac{B_n}{f_s} \qquad (8\text{--}28)$$

which provides a basis for comparing systems operating under different conditions. Re-arranging the right-hand side of Eq. (8–28),

$$\frac{S}{N}\frac{B_n}{f_s} = \frac{S}{\dfrac{N}{B_n}f_s}$$

and we obtain an alternative definition of R, namely

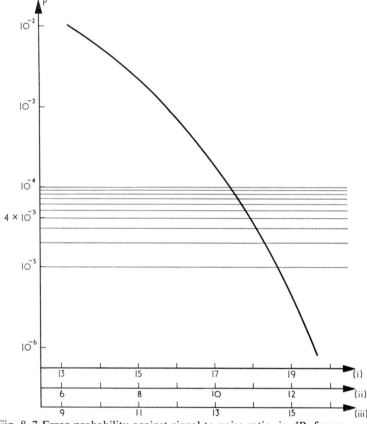

Fig. 8–7 Error probability against signal-to-noise ratio, in dB, for an on-off signal made up of full width rectangular pulses.

x—axes: (i) peak signal-to-mean noise power ratio, V^2/N;
           (ii) normalised signal-to-noise power ratio, and mean signal-to-noise power in the bit-rate bandwidth;
           (iii) mean signal-to-noise power ratio, V^2/N −4 dB.

normalised signal-to-noise ratio

$$= \frac{mean\ signal\ power}{noise\ power\ in\ the\ bit\text{-}rate\ bandwidth} \qquad (8\text{–}29)$$

and furthermore by substituting $f_s = 2f_0$ where f_0 is the Nyquist bandwidth, we have

$$\left\{ \frac{mean\ signal\ power}{noise\ power\ in\ the\ bit\text{-}rate\ bandwidth} \right\}$$

$$= \frac{1}{2} \left\{ \frac{mean\ signal\ power}{noise\ power\ in\ the\ Nyquist\ bandwidth} \right\} \qquad (8\text{–}30)$$

All three definitions which are to be found in common use are illustrated in Figs. 8–7 and 8–8 where the error probabilities of the on-off and polar forms of signalling are compared; it is assumed that k the pulse shape factor is unity and that optimum pulse shaping for zero crosstalk

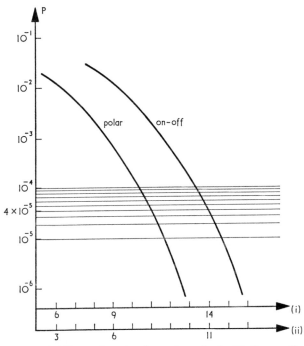

Fig. 8–8 Error probability against signal-to-noise ratio, in dB, for on-off and polar signals.

x—axes: (i) mean signal-to-noise ratio, or mean signal-to-noise power in Nyquist bandwidth, and
            (ii) mean signal-to-noise power in bit-rate bandwidth.

[8–2]

prior to detection is used; refer to Eq. (8–4). It may be shown that, for this filter, $B_n \approx f_o$ or $f_s/2$.

8–3 Error Probability and its Effect upon the Output Signal-to-noise Ratio of PCM and ∆M Systems

8–3.1 *PCM*

It has been shown in Chapter 6 that a PCM encoder quantizes the amplitude range of the analogue input into M states to allow conversion into binary code groups of n digits, where $M = 2^n - 1$. The output from a complementary decoder is inevitably a distorted version of the original signal, due to the process of quantization, but in the absence of digital errors this is the only source of noise in the output—interference accompanying the signal during transmission will have been removed by the decision detector. When the fluctuation noise is sufficiently large to cause a digital error in a code group, the decoder generates an incorrect amplitude level, thus giving rise to additional noise (or distortion), the degree of which depends upon the position in the code of the digit in error. Assuming linear quantization, an error in the least significant digit will cause the output to differ from the correct amplitude by the quantized step size Δv, and an error in the next significant digit would cause an output voltage difference of $2\Delta v$; in general, an error in the rth digit would give rise to an output error voltage $\Delta v\, 2^{(r-1)}$. If all digits in a code group of n digits are equally likely to be in error with probability P_e, and only one error will occur in the code group, the mean square noise voltage N_e is approximately

$$\sum_{r=1}^{n} \frac{P_e}{n} \{\Delta v 2^{(r-1)}\}^2$$

i.e.

$$N_e = P_e \frac{(\Delta v)^2}{n} \sum_{r=1}^{n} 2^{2(r-1)} \tag{8–31}$$

If we consider the analogue signal to be a sine wave of peak amplitude V and occupying the whole of the available amplitude range, then $2V = M\, \Delta v$ and the mean square signal power S is

$$S = \frac{M^2 (\Delta v)^2}{8} \tag{8–32}$$

Combining Eqs. (8–31) and (8–32) the mean square output signal-to-noise ratio, considering fluctuation noise only and not quantization noise, is

$$\frac{S}{N_e} = \frac{nM^2}{8\,P_e \displaystyle\sum_{r=1}^{n} 2^{2(r-1)}} \tag{8-33}$$

This equation is only valid for low error rates, say less than 1 in 100 digits, $P_e < 10^{-2}$; otherwise the signal component cannot be identified separately, as in Eq. (8–32), and hence the signal-to-noise ratio becomes the signal plus noise-to-noise ratio.

As a numerical illustration of Eq. (8–33) when $P_e = 10^{-3}$ and $n = 7$ ($M = 127$), the signal-to-noise ratio is 34 dB. The dashed curves of Fig. 8–9 show the error probability and output signal-to-noise ratio for on-off and polar signalling plotted against the mean signal-to-noise ratio at the input to the decision detector—i.e. the graph involves the use of Eq. (8–26), which is also illustrated in Fig. 8–8, and Eq. (8–33). The full curves of Fig. 8–9 include quantization noise which we see becomes the sole source of interference when the digital error probability becomes very small, 10^{-6} or less; in this example, a sine wave uniformly quantized into 127 levels is assumed which gives rise to a signal-to-quantizing noise ratio of 44 dB (see Table 7–1, page 140).

8–3.2 *ΔM*

The effects of digital errors on the output signal of a delta modulation system can be determined by representing the received re-shaped pulse sequence as an error-free pattern plus a randomly spaced series of rectangular pulses of amplitudes $+2V$ or $-2V$, the latter depending upon whether a 1 or a 0 is incorrectly received. A typical sequence is shown in Fig. 8–10; it is assumed that the error-free pattern, which is identical to the orginal transmitted signal, and the randomly spaced error pulses are received simultaneously.

Let the probability of an error be P_e, i.e. error pulses occur on average once every $1/P_e$ digits; half of the errors will give rise to $+2V$ pulses in waveform (iii) of Fig. 8–10, and the other half to $-2V$ pulses. Using the result of Tutorial Exercise III–4, page 281, the *two-sided* power density of the error pulse sequence is given by

$$S(f) = P_e f_s |G(j\omega)|^2$$

where

$$|G(j\omega)|^2 = \left[\frac{2V}{f_s}\frac{\sin(\omega/2f_s)}{(\omega/2f_s)}\right]^2$$

$$[8\text{–}3]$$

for the rectangular pulses. The *one-sided* error power density of the output from the RC integrator is

$$2 P_e f_s \, |G(j\omega)|^2 \, \frac{f_0^2}{f_0^2 + f^2}$$

and the output noise N_e from a band-pass filter having cut-off frequencies $f_{m1} = 300$ Hz and $f_{m2} = 3400$ Hz is given by

$$N_e = 2 P_e f_s f_0^2 \int_{f_{m1}}^{f_{m2}} |G(j\omega)|^2 \, \frac{1}{f_0^2 + f^2} \, df$$

and since $f_s \gg f_{m2}$ we may assume that $|G(j\omega)|^2 \approx 4 \, V^2/f_s^2$, hence

$$N_e \approx \frac{8 P_e V^2 f_0^2}{f_s} \int_{300}^{3400} \frac{1}{f_0^2 + f^2} \, df$$

$$= \frac{8 P_e V^2 f_0}{f_s} \left[\tan^{-1}\left(\frac{3400}{f_0}\right) - \tan^{-1}\left(\frac{300}{f_0}\right) \right] \qquad (8\text{–}34)$$

It follows that the mean signal-to-fluctuation noise power ratio corresponding to a sine wave of peak amplitude V is

$$\frac{S}{N_e} \approx \frac{f_s}{16 P_e f_0} \left[\tan^{-1}\left(\frac{3400}{f_0}\right) - \tan^{-1}\left(\frac{300}{f_0}\right) \right]^{-1} \qquad (8\text{–}35)^*$$

For example, when $f_0 = 150$ Hz and $f_s = 56$ kilobits/second, an error probability $P_e = 10^{-3}$ would result in $S/N_e = 48$ dB—this ratio does not include quantization noise.

A similar expression to Eq. (8–35) can be obtained for a basic ΔM system employing the double integration network of Fig. 7–17, page 169. Using Eq. (7–52) the output noise power N_e due to digital errors is

$$N_e = 2 P_e f_s \int_{f_{m1}}^{f_{m2}} |G(j\omega)|^2 \, \frac{f_2^2 f_1^2}{f^2(f^2 + f_2^2)} \, df$$

$$\approx \frac{8 P_e V^2 f_1^2 f_2^2}{f_s} \int_{f_{m1}}^{f_{m2}} \frac{1}{f^2(f^2 + f_2^2)} \, df$$

---

* This equation assumes that the mean output signal power is $V^2/2$. For a signal amplitude consistent with the conditions of non-overloading, Eq. (72–3) page 151 must be used and hence

$$\frac{S}{N_e} \approx \frac{f_s}{16 P_e (1 + \omega_m^2 T^2) f_0} \left[\tan^{-1}\left(\frac{3400}{f_0}\right) - \tan^{-1}\left(\frac{300}{f_0}\right) \right]^{-1}$$

Hence the signal-to-noise power ratio is given by

$$\frac{S}{N_e} \approx \frac{f_s}{16f_1{}^2 P_e}\left\{\left(\frac{1}{f_{m1}} - \frac{1}{f_{m2}}\right) - \frac{1}{f_2}\left[\tan^{-1}\left(\frac{f_{m2}}{f_2}\right) - \tan^{-1}\left(\frac{f_{m1}}{f_2}\right)\right]\right\}^{-1}$$

$$(8\text{-}36)$$

This is illustrated in Fig. 8–9 together with a plot of the output signal-to-noise ratio against the input ratio to the pulse regenerator. As we have seen for PCM, Eq. (8–35) and (8–36) only apply when the error probability is small, say $P_e < 10^{-2}$. The continuous curve for ΔM assumes that the digital signal is in polar form and made up of full-width rectangular pulses. A system employing double rather than single integration

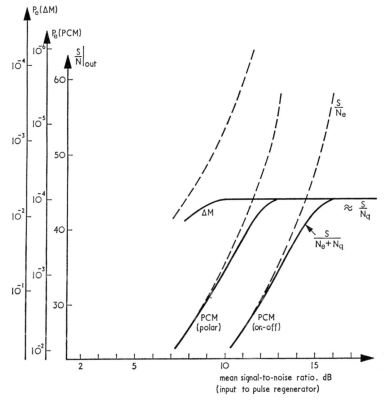

Fig. 8–9 Mean output signal-to-total noise power ratio plotted against the signal-to-noise power ratio at the input to the decision detector in the pulse regenerator.

PCM: sine wave signal uniformly quantized into 127 levels.

ΔM: basic system employing double integration and designed for speech.

(Broken lines represent S/N_e.)

[8–3]

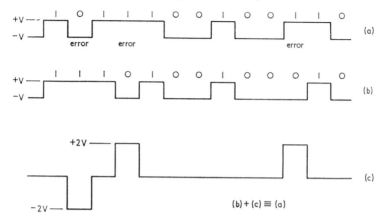

Fig. 8–10 Delta modulation.

(*a*) Received pulse sequence containing errors.
(*b*) Error-free sequence ≡ original transmitted sequence.
(*c*) Error-pulse sequence.

is illustrated simply for the convenience of comparing ΔM with PCM, which, in these particular examples, have the same signal-to-quantizing noise ratios. The important point to note is that fluctuation noise has a less disturbing effect on ΔM than it does on PCM; the effects of digital errors become significant when the error probability is 10^{-4} in PCM, whereas the corresponding figure for ΔM is 10^{-2}.

8–4 Spectra of Random-sequence Digital Signals

The essential requirement of a decision detection process, described in section 8-2, or a regenerative repeater is the synchronisation of the locally generated sampling pulses to the incoming digital signal. The synchronising information is an inherent characteristic of the incoming signal, but the question that arises is: what form does it take and how may it be extracted? It is shown in Appendix III that certain types of signals can exhibit a line spectrum and therefore, if one of the lines is at the digit repetition frequency, the simplest method of synchronisation is achieved by shock exciting a *LC* circuit with the digital signal and using the resulting sine wave to generate the sampling pulses. Before considering alternative methods of synchronisation for when a line spectrum does not exist, let us examine the spectra of the more common forms of signalling.

The general analysis is undertaken in Appendix III; re-stating Eqs. (III–37) and (III–38), the power contained in the line spectrum is given by

$$S\bigg|_{\text{line}} = \overline{[\text{av}\{^k x(t)\}]^2} = m_1{}^2 f_s{}^2 G^2(0) + 2m_1{}^2 f_s{}^2 \sum_{r=1}^{\infty} |G(jr\omega_s)|^2 \quad \text{watts}$$

(8–37)

(see also Eq. I–31 page 241), and the *one-sided* power density of the continuous spectrum as

$$S(f)\bigg|_{\text{continuous}} \approx 2f_s |G(j\omega)|^2 \{R(0) - m_1{}^2$$

$$+ 2 \sum_{q=1}^{\infty} [R(q) - m_1{}^2] \cos\omega q T_s\} \quad \text{watts/Hz}$$

(8–38)

Note the factor 2 difference between Eqs. (8–38) and (III–38)—the latter representing the *two*-sided spectrum.

The determination of the statistical constants m_1, m_2, $R(0)$ and $R(q)$ requires a knowledge of the coefficient a_n; refer to Appendix III, which defines a specific form of signalling. In the following analysis the values of a_n are chosen so that the same overall pulse amplitude range is available for all systems; for example, in an on-off system $a_n = +1$ and 0, in the polar system $a_n = +\frac{1}{2}$ and $-\frac{1}{2}$, in the bipolar system $a_n = +\frac{1}{2}$, 0 and $-\frac{1}{2}$, and so on. Normalisation of the amplitude range in preference to an equal mean power for all systems is chosen for two reasons: firstly, in line systems there is little point in choosing equal mean power as the normalisation criterion, since the power level is very low. Secondly, in radio transmission, FM and PM are invariably used for digital data and we have seen in previous chapters that the mean power is independent of the degree of modulation; a more important consideration is the frequency or phase deviation and hence the occupied bandwidth, which are directly proportional to the amplitude range of the modulating signal. Thus with the method of normalisation adopted it is possible to compare the various forms of signalling on a bandwidth basis, with all systems having the same available peak frequency or phase deviation.

8–4.1 *On-Off Signalling*

$$a_n = 1, \text{ with probability } p$$
$$= 0, \text{ with probability } 1-p$$

Therefore, $m_1 = \text{av } a_n = 1 \times p + 0(1-p) = p$

$$m_2 = \text{av } (a_n{}^2) = 1^2 \times p + 0^2(1-p) = p$$

and $\qquad R(0) = m_2 = p$ [8–4]

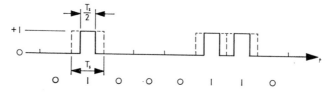

Fig. 8–11 A random on-off sequence of half-width rectangular pulses. (Broken lines represent full-width, NRZ pulses.)

Since there is zero interdigit correlation, av $(a_n a_m)$ = av a_n av a_m

and $\qquad R(q) = m_1^2 = p^2$

Substituting these constants into Eqs. (8–37) and (8–38) gives

$$S\bigg|_{\substack{\text{line,}\\ \text{on-off}}} = p^2 f_s^2 G^2(0) + 2p^2 f_s^2 \sum_{r=1}^{\infty} |G(jr\omega_s)|^2 \qquad (8\text{–}39)$$

and

$$S(f)\bigg|_{\substack{\text{continuous}\\ \text{on–off}}} \approx 2f_s |G(j\omega)|^2 [p(1-p)] \qquad (8\text{–}40)$$

Let us consider specific pulse shapes.

Example 8–2. Half-width rectangular pulses with equal probability of 1 and 0, i.e.

$$t_p = T_s/2 \text{ and } p = \tfrac{1}{2}$$

$$|G(j\omega)|^2 = \left[t_p \frac{\sin(\omega t_p/2)}{(\omega t_p/2)} \right]^2 = \left[\frac{T_s}{2} \frac{\sin(\pi f T_s/2)}{(\pi f T_s/2)} \right]^2$$

and $\qquad |G(jr\omega_s)|^2 = \left[\dfrac{T_s}{2} \dfrac{\sin(r\pi/2)}{(r\pi/2)} \right]^2$

Hence

$$S\bigg|_{\text{line}} = \frac{1}{16} + \frac{1}{8} \sum_{r=1}^{\infty} \left[\frac{\sin(\pi r/2)}{(\pi r/2)} \right]^2 \text{ watts}$$

and $\qquad S(f)\bigg|_{\text{continuous}} \approx \dfrac{T_s}{8} \left[\dfrac{\sin(\pi f T_s/2)}{(\pi f T_s/2)} \right]^2 \text{ watts/Hz}$

Note that the line spectrum vanishes at any harmonic frequency $r f_s$ for which $G(jr\omega_s) = 0$ (see Fig. 8–12 (a)).

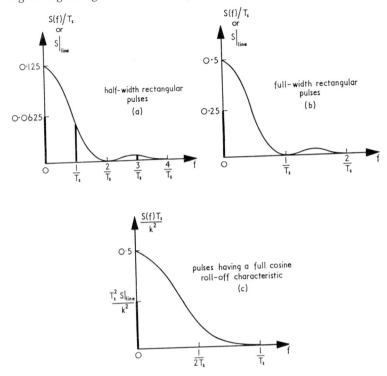

Fig. 8–12 Continuous and line spectra of random sequence on-off binary signals with specific pulse shapes.

Example 8–3. Full-width rectangular pulses (non-return to zero, NRZ) $t_p = T_s$ and equal probability of 1 and 0.

$$G(jr\omega_s) = T_s \frac{\sin r\pi}{r\pi}$$

and is zero for all values of r, except $r = 0$ the d.c. component, which means that a line spectrum does not exist. This may, at first sight, appear to be a remarkable result and one which the reader may treat with some suspicion. Any doubts regarding its accuracy can be dispelled by drawing a sketch similar to Fig. III–4 and examining the excitation of a LC circuit resonant at the digit rate f_s by a random sequence of full-width pulses; the reader will verify that succeeding pulses have a destructive effect upon any oscillation that may have started previously. From Eq. (8–40) the continuous spectrum is defined as

$$S(f)\bigg|_{\text{continuous}} \approx \frac{T_s}{2}\left[\frac{\sin(\pi f T_s)}{\pi f T_s}\right]^2 \text{ watts/Hz}$$

and the d.c. component from Eq. (8–39) is 0·25 (see Fig. 8–12 (b)).   [8–4]

Example 8–4. Pulses having a full cosine roll-off spectrum.

From Eq. (8–1)

$$G(j\omega) = k\cos^2\frac{\pi f}{4f_0} \quad \text{volt-seconds}, \quad 0 < f < 2f_0$$

and
$$f_s = 2f_0$$

Using Eq. (8–39), the line spectrum is

$$S\Big|_{\text{line}} = \frac{k^2}{4T_s^2}\Big|_{\text{d.c.}} + \frac{k^2}{2T_s^2}\cos^4\left(\frac{\pi r f_s}{4f_0}\right)$$

and the second term is equal to zero; hence there is a d.c. component only. The power density of the continuous spectrum is

$$S(f)\Big|_{\text{continuous}} \approx \frac{k^2}{2T_s}\cos^4\left(\frac{\pi f}{2f_s}\right), \quad 0 < f < f_s$$

which is illustrated in Fig. 8–12 (c).

8–4.2 *Polar Signalling*

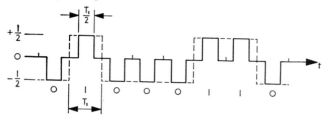

Fig. 8–13 A random polar sequence of half-width rectangular pulses. (Broken lines represent full-width, NRZ, pulses.)

$$a_n = \tfrac{1}{2}, \text{ with probability } p$$
$$= -\tfrac{1}{2}, \text{ with probability } 1-p$$

Therefore

$$m_1 = \text{av } a_n = (\tfrac{1}{2})p + (-\tfrac{1}{2})(1-p) = p-\tfrac{1}{2}$$
$$m_2 = \text{av}(a_n^2) = (\tfrac{1}{2})^2 p + (-\tfrac{1}{2})^2(1-p) = \tfrac{1}{4}$$

and
$$R(0) = m_2 = \tfrac{1}{4}.$$

Since there is zero interdigit correlation,

$$R(q) = m_1^2 = (p-\tfrac{1}{2})^2$$

Substituting these constants into Eqs. (8–37) and (8–38) gives

$$S\Big|_{\substack{\text{line,} \\ \text{polar}}} = (p-\tfrac{1}{2})^2 f_s^2 G^2(0) + 2(p-\tfrac{1}{2})^2 f_s^2 \sum_{r=1}^{\infty} |G(jr\omega_s)|^2 \qquad (8\text{–}41)$$

and

$$S(f)\Bigg|_{\substack{\text{continuous,}\\ \text{polar}}} \approx 2f_s |G(j\omega)|^2 [p(1-p)] \qquad (8\text{–}42)$$

Example 8–5. Half-width pulses and equal probability of the two states, i.e. $p = \frac{1}{2}$.

Firstly, $m_1 = 0$ and therefore a line spectrum does not exist.

$$|G(j\omega)|^2 = \left[\frac{T_s}{2} \frac{\sin(\pi f T_s/2)}{\pi f T_s/2}\right]^2$$

and therefore the one-sided continuous spectrum is defined by

$$S(f)\Bigg|_{\text{continuous}} \approx \frac{T_s}{8}\left[\frac{\sin(\pi f T_s/2)}{\pi f T_s/2}\right]^2$$

8–4.3 *Bipolar Signalling*

This is a pseudo-binary form of signalling in which a 0 is transmitted as no-pulse and the 1's are transmitted alternately as positive and negative pulses,

$$\left. \begin{array}{l} a_n = +\tfrac{1}{2} \\ = -\tfrac{1}{2} \\ = 0 \end{array} \right\} \text{ with probability } \left\{ \begin{array}{l} p/2 \\ p/2 \\ 1-p \end{array} \right.$$

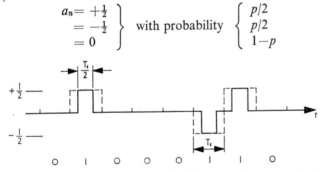

Fig. 8–14 A random bipolar sequence of half-width rectangular pulses. (Broken lines represent full-width, NRZ, pulses.)

Therefore

$$m_1 = \text{av } a_n = (+\tfrac{1}{2})\frac{p}{2} + 0(1-p) + (-\tfrac{1}{2})\frac{p}{2} = 0$$

$$m_2 = \text{av } (a_n{}^2) = (+\tfrac{1}{2})^2\frac{p}{2} + 0^2(1-p) + (-\tfrac{1}{2})^2\frac{p}{2} = \frac{p}{4}$$

and $\qquad R(0) = m_2 = \dfrac{p}{4}$

$$[8\text{–}4]$$

The evaluation of $R(q)$ must take into account the interdigit correlation which for this form of signalling is not zero. From the jth to the $(j+q)^{th}$ digit inclusive there are $(q+1)$ digits; the total number of possible binary patterns is 2^{q+1}. Any pattern with a 0 at the beginning or end would have a product $a_j a_{j+q} = 0$, and therefore we need only consider those patterns starting and ending in 1, of which there are 2^{q-1}. The interdigit autocorrelation function may be written as

$$R(q) = (+\tfrac{1}{4})\frac{A}{2^{q+1}} + (-\tfrac{1}{4})\frac{B}{2^{q+1}}$$

where A is the number of patterns starting and ending with $+\tfrac{1}{2}$ or $-\tfrac{1}{2}$ and B is the number of patterns starting with $+\tfrac{1}{2}$ and ending with $-\tfrac{1}{2}$, or vice versa.

When q = 1, $A = 0$ and $B = 1$

$\qquad$ q = 2, $A = 1$ and $B = 1$

in fact, when $q \geqslant 2$, $A = B$

and therefore

$$R(1) = (-\tfrac{1}{4})\frac{1}{2^{1+1}} = -\frac{1}{16}$$

and $\qquad\qquad R(q) = 0$ for $q \geqslant 2$.

The average value m_1 is zero and therefore a line spectrum does not exist. When there is equal probability of 1 and 0, i.e. $p = \tfrac{1}{2}$, the one-sided power density of the continuous spectrum is given by

$$S(f)\Big|_{\substack{\text{continuous,}\\ \text{bipolar}}} \approx 2f_s |G(j\omega)|^2 \left[\tfrac{1}{8} + 2(-\tfrac{1}{16})\cos\omega T_s\right]$$

$$= \frac{1}{4T_s}|G(j\omega)|^2(1-\cos\omega T_s) \qquad (8\text{--}43)$$

8–4.4 Duobinary Signalling[4]

Like the bipolar form of signalling, the duobinary technique can also be classified as pseudo-binary—0 is transmitted as no pulse and 1 as either a positive or negative pulse depending upon the previous sequence. The choice of polarity is most easily understood by considering the conversion of an on-off sequence into duobinary form. Briefly, if there are an odd number of 0s between the previous 1 and the next to be generated, the polarity of the latter is reversed; when the number of 0s is even the

4. The Duobinary Technique for High-Speed Data Transmission, A. Lender, *Trans. I.E.E.E.*, Vol. 82, pp. 214–18, May 1963.

polarity is unchanged (see Fig. 8–15). Unlike the bipolar form, a reversal from a positive to a negative pulse in adjacent time slots is impossible. The statistical constants may be defined as follows:

$$\left. \begin{array}{l} a_n = +\tfrac{1}{2} \\ \quad = -\tfrac{1}{2} \\ \quad = 0 \end{array} \right\} \quad \text{with probability} \quad \left\{ \begin{array}{l} p/2 \\ p/2 \\ 1-p \end{array} \right.$$

$$\left. \begin{array}{l} m_1 = \text{av } a_n = 0 \\ m_2 = \text{av } (a_n{}^2) = \dfrac{p}{4} \end{array} \right\} \quad \text{as for the bipolar form,}$$

and
$$R(0) = m_2 = \frac{p}{4}$$

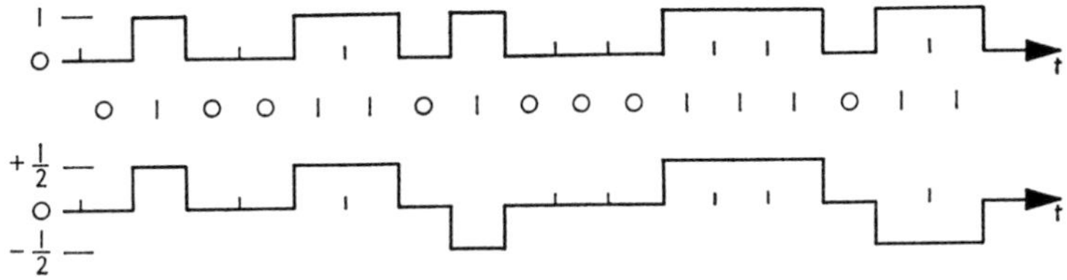

Fig. 8–15 A random on-off sequence of full-width, NRZ, rectangular pulses and the equivalent duobinary form.

$R(q)$ may be evaluated by a similar approach to that used for the bipolar signal. Therefore we have

$$R(q) = \text{av } (a_j a_{j+q}) = (+\tfrac{1}{4})\frac{A}{2^{q+1}} + (-\tfrac{1}{4})\frac{B}{2^{q+1}}$$

when
$$q = 1, \, A = 1 \text{ and } B = 0$$
$$q = 2, \, A = 1 \text{ and } B = 1$$
$$q = 3, \, A = 2 \text{ and } B = 2$$

in fact, when $q \geqslant 2$, $A = B$ and $R(q) = 0$. This leaves

$$R(1) = (+\tfrac{1}{4})\frac{1}{2^{1+1}} = +\frac{1}{16}$$

The average value, m_1, is zero and therefore a line spectrum does not exist. When there is equal probability of 1 and 0, $p = \tfrac{1}{2}$, and the one-sided power density of the continuous spectrum is

$$S(f)\bigg|_{\substack{\text{continuous,} \\ \text{duobinary}}} \approx 2f_s \, |G(j\omega)|^2\{\tfrac{1}{8} + 2(\tfrac{1}{16})(1 + \cos\omega T_s)\}$$

$$= \frac{1}{4T_s}|G(j\omega)|^2(1 + \cos\omega T_s) \qquad (8\text{–}44)$$

$$[8\text{–}4]$$

A comparison of the four methods of signalling is made in Fig. 8–16; the curves illustrate Eqs. (8–40), (8–42) — both for $p = \frac{1}{2}$ — (8–43) and (8–44). The line spectrum associated with the on-off form of signalling has been omitted from the figure.

In line systems extensive use is made of the bipolar form of signalling, the main advantage being the absence of power at d.c. and the very low frequencies, thus easing the bandwidth requirements of the repeater amplifiers. The duobinary technique is favoured for radio transmission

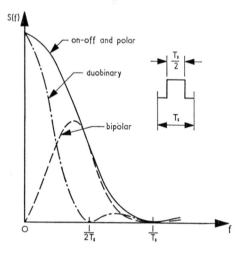

Fig. 8–16 Continuous spectra of random sequence half-width rectangular pulses with specific forms of signalling.

on account of it occupying a much narrower bandwidth than the other forms of signalling. It is to be seen that neither bipolar nor duobinary signals exhibit a line spectrum, and therefore the timing information necessary for correct sampling at the repeaters and final receiver cannot be extracted linearly with a simple LC resonant circuit. In general, a non-linear operation is required; one method is to convert the signal into the equivalent on-off form at the repeater or receiver, which may then be used to excite the LC circuit.

Tutorial Exercises

8–1 Show that the impulse response of a filter that is a matched filter for an unsymmetrical input pulse is a delayed replica of the pulse, but run backwards in time (see page 179).

8–2 Prove that the matched filter for a rectangular pulse in the presence of Gaussian-distributed noise has a sin x/x amplitude frequency characteristic.

8-3 A 1 μs triangular pulse of 1 V maximum amplitude in the presence of white noise of spectral density $1\cdot19 \times 10^{-8}$ V²/Hz is applied to a matched filter. Calculate the peak signal-to-mean noise power ratio at the output of the filter.

The same system is used to maximise the signal-to-noise ratio of a random on-off sequence of these pulses. The signal is decoded by inspecting each pulse (or no-pulse) when the signal-to-noise ratio is a maximum. The slicing level for detection is set at half the peak amplitude of the output pulse.

Estimate the average time between errors when the pulse repetition rate is 10^6 per second.

Assume

    (i) equal probability of pulse or no-pulse being transmitted, and
    (ii) that intersymbol interference can be ignored.

Ans: 17·4 dB; 10 ms.

8-4 A digital signal made up of rectangular pulses in polar form, equal positive and negative peak amplitudes representing the two binary states, is received in the presence of white band-limited noise. The mean signal-to-noise ratio at the input to a decision detector is 12·8 dB. Calculate the *minimum* average error rate.

Ans: 7 in 10^6.

8-5 When an analogue signal is transmitted in digital form the received reconstructed signal will be impaired due to *aliasing, quantization* and *noise pick-up*. Explain how these phenomena may be minimised. Are they in any way interrelated and, if so, how does it affect their minimisation?

8-6 A binary transmission of amplitude modulated on-off pulses in the presence of band-limited Gaussian noise is demodulated by an envelope detector. The mean signal-to-noise power ratio at the input to the detector is 16 dB.

The output is decoded by inspecting the pulse train to determine whether the resultant signal is above or below a certain threshold level.

Evaluate the threshold level, in terms of the r.m.s. noise voltage, to give the minimum probability of error.

Assume

    (i) equal probability of signal or no-signal,
    (ii) envelope of signal plus noise has a Gaussian distribution with first moment equal to the peak amplitude of the signal, and
    (iii) envelope of noise alone has a Rayleigh distribution

$$p(r) = \frac{r}{N} \exp\left(-\frac{r^2}{2N}\right)$$

where N is the mean noise power.

Hint: Plot error probability against threshold level for signal plus noise, and for noise alone.

Ans: threshold level $\approx 4\cdot75 \times r.m.s.$ noise voltage.

8–7 A random sequence of binary digits, referred to as a *partial response* signal, is made up of two interleaved bipolar pulse sequences, see Reference 5.

Show that the continuous power spectrum of an ensemble of these pulses is given by

$$S(f) \approx 2f_s \left| G(j\omega) \right|^2 \left(\frac{p}{4} - \frac{1}{8} \cos 2\omega T_s \right) \text{ watts/Hz}$$

(Let $a_n = \pm\frac{1}{2}$, as in Section 8–4.3.)

5. Some Practical Aspects of Digital Transmission, J. R. Pierce, *I.E.E.E. Spectrum*, Vol. 5, No. 8, pp. 63–70, November 1968.

Appendix I
TIME AND FREQUENCY
DOMAIN ANALYSIS

I–1 Introduction

The communications engineer is concerned with the processing and transmission of time-varying signals conveying information. He requires to know the bandwidths occupied by the signals in order to answer such questions as, how many independent signals may be accommodated in a multi-channel system of fixed bandwidth allocation, and will interference be experienced by other signals occupying adjacent channels? Ideally, the characteristics of a signal should be definable in terms of frequency by a relatively simple mathematical expression. This is seldom possible, particularly for analogue signals. Consider speech. The bandwidth is arbitrarily defined as 300–3400 Hz by subjective analysis of quality and intelligibility; the actual way in which the bandwidth is occupied depends upon many factors of the human voice and cannot be expressed by simple algebraic definition. Similarly, the instantaneous amplitude of speech as a simple time-dependent function eludes definition. Instead the signal can only be defined in terms of its statistical properties (see Chapter 7).

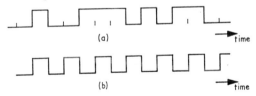

Fig. I–1

(a) Random sequence pulse train.
(b) Periodic pulse train (alternate marks and spaces).

Next, consider a digital signal of on-off binary pulses; if the signal is conveying information, the pulse sequence is random in nature (see Fig. I–1 (a)), and the time dependency can only be defined in statistical terms. We may now ask the question, what is the bandwidth occupied by such a signal, and, since it cannot be arbitrarily defined by subjective

[I–1]

analysis, as for speech, can the signal representation in the frequency domain be obtained from a knowledge of the time-dependence characteristics? The answer to this question is 'yes', but the analytical tools must be developed by firstly considering a binary signal that is periodic in every respect (see Fig. I–1 (*b*)). It must also be assumed, for mathematical convenience and ease of handling, that the pulse train commenced at time $t = -\infty$ and continues to $t = +\infty$. It will be appreciated that such a signal conveys zero information, since its state at any instant in time can be deduced with complete certainty from past events. Although this may seem to place a prohibitive restriction on the usefulness of the analysis, it will be seen that the duality of the time and frequency domains for this periodic pulse train has many applications.

I–2 Fourier Series Representation of a Periodic Signal
I–2.1 *Trigonometric Form*

Any periodic function $f(a)$ can be represented as a series

$$f(a) = a_0 + a_1\cos a + a_2\cos 2a + a_3\cos 3a + \ldots$$
$$+ b_1\sin a + b_2\sin 2a + b_3\sin 3a + \ldots \qquad (I–1)$$

where the period is 2π and $f(a) = f(a+2\pi)$. (Note that the variable a is in radians.)

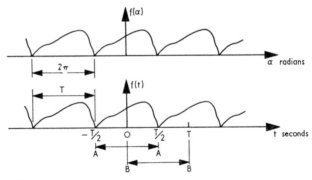

Fig. I–2 The equivalence of the periodic functions $f(a)$ and $f(t)$.

For our purposes the periodic signals will be time dependent with a period T. Therefore, the periodic function $f(t)$ can be defined as

$$f(t) = a_0 + \sum_{n=1}^{\infty} a_n\cos\frac{2\pi nt}{T} + \sum_{n=1}^{\infty} b_n\sin\frac{2\pi nt}{T} \qquad (I–2)$$

since $a/2\pi = t/T$ (see Fig. I–2).

The periodicity (or frequency) of the function can also be defined as

$$f = \frac{1}{T} \text{ periods/second} \tag{I–3}$$

or as an angular frequency $\omega = 2\pi/T$ radians/second.

This is the Fourier series representation of a periodic function and is valid provided that $\int |f(t)|\, dt$ over a complete period is finite and $f(t)$ is piecewise continuous. These conditions are satisfied for the various periodic functions considered in this text and, in general, for practically all physically realizable functions.

Let us consider the coefficients a_0, a_n, and b_n. If, for example, $f(t)$ defines the periodic variations of current through a circuit, then a_0 will represent the d.c. component, and the coefficients a_n and b_n will define the amplitudes of the harmonically related a.c. components. The trigonometrical integrals listed at the bottom of the page are required for the evaluation of these coefficients for a specific time-varying function.

First, to find a_0, integrate both sides of Eq. (I–2) with respect to t over a complete period, say AA (see Fig. I–2). Then

$$\int_{-T/2}^{T/2} f(t)\, dt = \int_{-T/2}^{T/2} a_0\, dt + \int_{-T/2}^{T/2} \left[\sum_{n=1}^{\infty} (a_n \cos n\omega t + b_n \sin n\omega t) \right] dt$$

$$= 2\pi a_0$$

or
$$a_0 = \frac{1}{T} \int_{-T/2}^{T/2} f(t)\, dt \tag{I–4a}$$

A similar result is obtained if the period over which $f(t)$ is integrated is taken as BB (see Fig. I–2). The limits of integration are changed and the coefficient a_0 is given by

$$a_0 = \frac{1}{T} \int_{0}^{T} f(t)\, dt \tag{I–4b}$$

$$\int_{-T/2}^{T/2} \cos \omega t\, dt = 0 \qquad \int_{-T/2}^{T/2} \cos \omega_1 t \cos \omega_2 t\, dt = 0, \ \omega_1 \neq \omega_2$$

$$\int_{-T/2}^{T/2} \sin \omega t\, dt = 0 \qquad \int_{-T/2}^{T/2} \sin \omega_1 t \sin \omega_2 t\, dt = 0, \ \omega_1 \neq \omega_2$$

$$\int_{-T/2}^{T/2} \cos \omega_1 t \sin \omega_1 t\, dt = 0$$

$$\int_{-T/2}^{T/2} \sin^2 \omega t\, dt = \int_{-T/2}^{T/2} \cos^2 \omega t\, dt = \frac{T}{2} \tag{I–2}$$

To find a_n, multiply both sides of Eq. (I–2) by cos $n\omega t$ and integrate over the period AA. On the right-hand side all the integrals vanish except the one involving $a_n\cos^2 n\omega t$.

Therefore
$$\int_{-T/2}^{T/2} f(t)\cos n\omega t\, dt = \int_{-T/2}^{T/2} a_n\cos^2 n\omega t\, dt = a_n T/2$$

Hence
$$a_n = \frac{2}{T}\int_{-T/2}^{T/2} f(t)\,\cos n\omega t\, dt \qquad\qquad\text{(I–5a)}$$

or
$$a_n = \frac{2}{T}\int_0^T f(t)\,\cos n\omega t\, dt \qquad\qquad\text{(I–5b)}$$

Similarly, multiplying Eq. (I–2) by sin $n\omega t$ and integrating gives

$$b_n = \frac{2}{T}\int_{-T/2}^{T/2} f(t)\,\sin n\omega t\, dt \qquad\qquad\text{(I–6a)}$$

or
$$b_n = \frac{2}{T}\int_0^T f(t)\,\sin n\omega t\, dt \qquad\qquad\text{(I–6b)}$$

Example I–1. Evaluate the Fourier coefficients for a periodic waveform of rectangular pulses having a period T and pulse duration t_p.

It is necessary to define $f(t)$ over a period and this will involve an arbitrary choice of time origin; let this be taken at the centre of the pulse so that

$$f(t) = V, \quad -t_p/2 < t < t_p/2$$
$$f(t) = 0, \quad t > t_p/2 \quad \left.\begin{array}{l} \text{for the period centred} \\ \text{on the time origin} \end{array}\right.$$
$$f(t) = 0, \quad t < -t_p/2 \quad \text{(see Fig. I–3).}$$

Using Eq. (I–4a), $a_0 = \dfrac{1}{T}\displaystyle\int_{-T/2}^{T/2} V\, dt = \dfrac{V t_p}{T}$

i.e. the average value of the waveform.

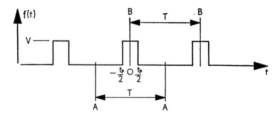

Fig. I–3 Periodic train of rectangular pulses, amplitude V, pulse duration t_p and period T.

Solving for a_n using Eq. (I–5a),

$$a_n = \frac{2}{T}\int_{-t_p/2}^{t_p/2} V\cos n\omega t \, dt$$

$$= \frac{2V}{n\pi} \sin\left(\frac{n\pi t_p}{T}\right)$$

Similarly, from Eq. (I–6a),

$$b_n = \frac{2}{T}\int_{-t_p/2}^{t_p/2} V\sin n\omega t \, dt = 0$$

therefore

$$f(t) = \frac{Vt_p}{T} + \sum_{n=1}^{\infty} \frac{2V}{n\pi} \sin\left(\frac{n\pi t_p}{T}\right) \cos n\omega t \tag{I–7}$$

It can be seen that the infinite series of sine terms has vanished in this particular solution because b_n is zero.

Tutorial Exercise I–1. Obtain the same result as Eq. (1–7) by evaluating the coefficients over the period *BB* in place of *AA* (see Fig. I–3), but keeping the origin $t = 0$ at the centre of the pulse.

In general, for any periodic function, if the time origin can be chosen so that $f(t) = f(-t)$ the coefficients b_1, b_2 . . . b_n will be zero and the function is said to have *even symmetry* about the time origin.

If $f(t) = -f(-t)$ the coefficients a_1, a_2 . . . a_n are zero and *odd symmetry* is said to exist (see Fig. I–4).

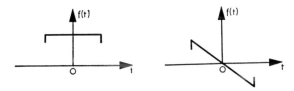

Fig. I–4 Examples of even and odd symmetry.

Therefore, when a periodic function is expressed as a Fourier series, the latter can be simplified to a single infinite series involving either $\cos n\omega t$ terms or $\sin n\omega t$ terms if the time origin can be chosen for symmetry to exist.

Referring to Example I–1, it was possible to arrange for even symmetry by positioning the time origin at the centre of a pulse. If, for example, the leading edge of a pulse had been chosen as the time origin,

[I–2]

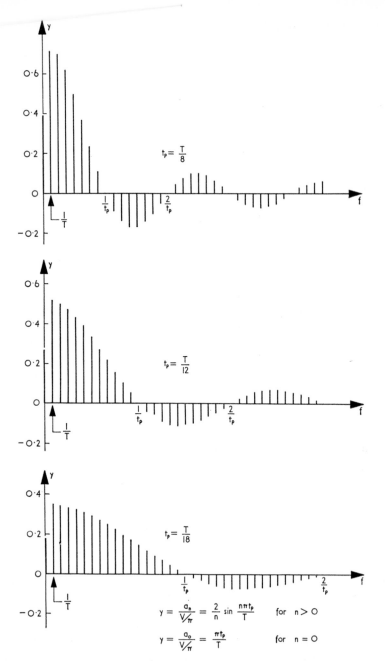

Fig. I–5 Amplitude-frequency spectra of a periodic train of rectangular pulses for $t_p = T/8$, $T/12$ and $T/18$.

then neither even nor odd symmetry alone would have existed and the resulting Fourier series would be made up of cos $n\omega t$ and sin $n\omega t$ terms.

Let us consider the example further, and the significance of Eq. (I–7). The first term Vt_p/T defines the average value of the periodic signal. The infinite series of harmonically related cosine terms means that the time-varying waveform is represented by a discrete spectrum in the frequency domain.* The peak amplitude of the nth component (or harmonic) is given by $\{2V \sin(n\pi t_p/T)\}/n\pi$ and the cos $n\omega t$ factor will determine the instantaneous amplitude of the nth component. It is often helpful, in order to appreciate the frequency spectrum of a particular signal, to present the Fourier series graphically; this can be achieved in two forms. Firstly, Fig. I–5 shows a plot of amplitude against frequency for $t = 0$; the positive and negative sign change due to the sin $(n\pi t_p/T)$ term is included. (Note, by inspecting Eq. (I–7), a knowledge of the pulse duration t_p relative to the period time T is required; three conditions are considered in Fig. I–5, $t_p = T/8$, $T/12$ and $T/18$.)

The second representation (Fig. I–6) displays the modulus of the amplitude against frequency, and the positive and negative sign changes are shown separately, again for the condition $t = 0$. The spectral components are defined in terms of amplitude and phase. Generally, when one refers to the phase of a waveform, or more accurately to the phase difference between the waveform and some reference, it is assumed that their frequencies are the same. In the present application we are referring to the amplitudes of components at different frequencies, but at a given instant in time $t = 0$. It is therefore more meaningful to talk about the instantaneous phases of these components rather than simply the phases. Note, that if a large number of the spectral components were summed, taking into account the instantaneous phases, the resultant would approximate closely to the amplitude of $f(t)$ at $t = 0$. The word 'approximate' is used, since the exact value of $f(t)$ could only be obtained by summing an infinite number of components.

In the above analysis of a periodic rectangular pulse train it has been possible to define a period with even symmetry about the time origin. Next, consider a periodic function for which symmetry about the time origin cannot be established. The Fourier series will contain an infinite number of cosine terms and sine terms, and a plot of the line spectrum is not as straightforward as in the previous example, since for a given

---

* The concept of frequency is fundamental to electrical subjects in general, and for this reason is often taken for granted. In this instance it is the substitution of Eq. (I–3) into the Fourier series that enables the time and frequency duality to be established.

value of n there are two terms in phase quadrature which define the amplitude and instantaneous phase of the particular harmonic. The modulus of the amplitude of the spectral component is given by $\sqrt{a_n^2 + b_n^2}$ and the instantaneous phase as $\tan^{-1}(b_n/a_n)$. Although these are simple mathematical statements, their numerical evaluation is tedious and lengthy. Some of the arithmetic can be avoided by using a different

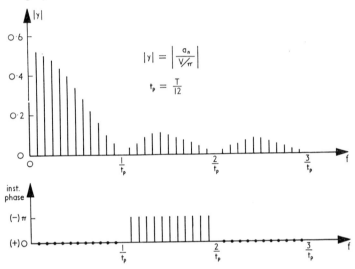

Fig. I–6 Line spectrum shown in Fig. I–5 for $t_p = T/12$, but with the amplitude and phase information plotted separately. (Figs. I–5 and I–6 represent Eq. (I–7) for the condition $t = 0$.)

form for the general Fourier series representation, which involves only a single infinite series regardless of whether the periodic function exhibits symmetry or not.

I–2.2 *Complex Form*
This method of defining the Fourier series makes use of the exponential function $\exp(j\omega t)$. Consider the cosine and sine terms of the nth component being combined as follows:

$$a_n\cos n\omega t + b_n\sin n\omega t = c_n\cos\phi_n\cos n\omega t + c_n\sin\phi_n\sin n\omega t$$

$$\text{where } c_n = (a_n^2 + b_n^2)^{\frac{1}{2}}$$

$$\text{and } \phi_n = \tan^{-1}(b_n/a_n)$$

Also $c_n\cos\phi_n\cos n\omega t + c_n\sin\phi_n\sin n\omega t = c_n\cos(n\omega t - \phi_n)$

which may be expressed as

$$\tfrac{1}{2}c_n\exp\{j(n\omega t - \phi_n)\} + \tfrac{1}{2}c_n\exp\{-j(n\omega t - \phi_n)\}$$

Therefore the original function $f(t)$ can be written as

$$f(t) = \sum_{n=-\infty}^{\infty} d_n \exp(jn\omega t) \qquad (I-8)$$

where
$$d_n = \tfrac{1}{2}c_n \exp(-j\phi_n) \qquad (I-9a)$$

and
$$d_{-n} = \tfrac{1}{2}c_n \exp(+j\phi_n) \qquad (I-9b)$$

Before proceeding, there are two features of Eq. (I-8) that require explanation. Firstly, the summation from $-\infty$ to $+\infty$ means that the line spectrum will contain components at positive *and* negative frequencies. The latter have no physical meaning and their inclusion is due to the mathematical model used for representing the Fourier series. At this stage, let us just note this phenomenon; the full significance will be considered later. Secondly, it would appear that $f(t)$ is a complex function. For our purposes $f(t)$ is time dependent and is always real, consequently a_n, b_n, c_n and ϕ_n are all real quantities. But according to Eq. (I-9) the coefficient d_n is complex, therefore the condition

$$d_n = d_{-n}^{\,*}$$

must apply for $f(t)$ to be a real function. This means that the imaginary part of the nth positive frequency component is equal but of opposite sign to the imaginary part of the negative frequency complement.

For the application of Eq. (I-8) it remains to solve the coefficient d_n. Using Eqs. (I-5a) and (I-6a) we can write

$$a_n - jb_n = \frac{2}{T}\int_{-T/2}^{T/2} f(t)\tfrac{1}{2}\{\exp(jn\omega t)+\exp(-jn\omega t)\}\ dt$$

$$- \frac{2}{T}\int_{-T/2}^{T/2} f(t)\tfrac{1}{2}\{\exp(jn\omega t)-\exp(-jn\omega t)\}\ dt$$

$$= c_n(\cos\phi_n - j\sin\phi_n)$$

$$= c_n \exp(-j\phi_n)$$

Therefore,
$$\frac{2}{T}\int_{-T/2}^{T/2} f(t)\exp(-jn\omega t)\ dt = c_n\exp(-j\phi_n)$$

and by substitution using Eq. (I-9),

$$d_n = \frac{1}{T}\int_{-T/2}^{T/2} f(t)\exp(-jn\omega t)\ dt \qquad (I-10)$$

[I-2]

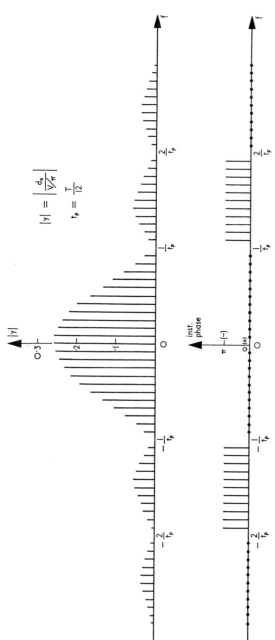

Fig. I–7 Two-sided representation of the spectrum corresponding to a periodic train of rectangular pulses for which the time origin coincides with the centre of a pulse (see Example I–2, Eq. (I–12)).

Eqs. (I–8) and (I–10) provide the second method of representing a time varying periodic function as a Fourier series defining the frequency spectrum.

Example I–2. Take the periodic train of rectangular pulses defined in Fig. I–3. When the time origin is arranged for even symmetry, the coefficient d_n is given by

$$d_n = \frac{1}{T}\int_{-t_p/2}^{t_p/2} V\exp(-jn\omega t)dt$$

$$= \frac{V}{T}\left[\frac{\exp(-jn\omega t)}{-jn\omega}\right]_{-t_p/2}^{t_p/2}$$

$$= \frac{V}{T}\frac{2\sin(n\omega t_p/2)}{n\omega}$$

$$= \frac{V}{n\pi}\sin(n\pi t_p/T) \qquad (I\text{--}11)$$

Using Eq. (I–8),

$$f(t) = \sum_{n=-\infty}^{\infty} \frac{V}{n\pi}\sin(n\pi t_p/T)\exp(jn\omega t) \qquad (I\text{--}12)$$

The average value of $f(t)$, given by the coefficient d_0, must be evaluated separately by substituting $n = 0$ in the integral above.

This gives $d_0 = Vt_p/T$

The spectral components are shown graphically in Fig. I–7 as the modulus of the amplitude of each component and its instantaneous phase, for $t = 0$ and $t_p/T = 12$. A comparison of the two-sided spectrum, as it is called, and the one-sided spectrum of Fig. I–6 is dealt with in Section I–3.

Tutorial Exercise I–2. Show that the same result as Eq. (I–12) can be obtained by defining $f(t)$ over the period BB in place of AA (see Fig. I–3).

It was seen in the first method of Fourier series analysis that if the time origin could be positioned for either even or odd symmetry to exist the Fourier representation reduces to a single series involving either a_n or b_n. If the alternative exponential form is used, the coefficient d_n is real when symmetry about the time origin can be established, but complex when symmetry does not exist; the following example serves as an illustration.

Example I–3. A periodic train of rectangular pulses is defined with the
[I–2]

time origin coinciding with the leading edge of a pulse (see Fig. I–8). Then over the period BB

$$f(t) = V, \quad 0 < t < t_p$$
$$f(t) = 0, \quad t > t_p$$

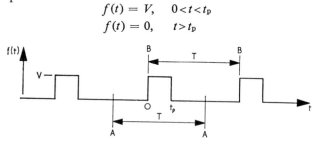

Fig. I–8 Rectangular pulse train with the time origin chosen to coincide with the leading edge of a pulse.

The Fourier coefficient d_n is

$$d_n = \frac{1}{T} \int_0^{t_p} V \exp(-jn\omega t) dt$$

$$= \frac{V}{n\pi} \sin(n\pi t_p/T) \exp(-jn\pi t_p/T) \qquad \text{(I–13)}$$

Therefore, $\qquad f(t) = \sum_{n=-\infty}^{\infty} \frac{V}{n\pi} \sin(n\pi t_p/T) \exp(-jn\pi t_p/T) \exp(jn\omega t) \quad \text{(I–14)}$

The modulus of d_n and the instantaneous phase of each component, for $t_p = T/12$, is displayed graphically in Fig. I–9. A comparison of Fig. I–7 and Fig. I–9 shows that the amplitude spectra are the same and that they only differ in respect of their *phase* spectra. This, of course, is to be expected, since the arbitrary positioning of the time origin will determine the relative phases of the components at some instant in time, whereas the moduli of their amplitudes are determined solely by the characteristics of the periodic waveform and independent of choice of time origin.

I–3 Comparison of One-sided and Two-sided Fourier Representations

It has been shown in section I–2 that the one-sided and two-sided spectra apply to the trigonometric form and complex form of Fourier series respectively. The latter is the preferred form, because it involves only one coefficient d_n, in place of a_n and b_n for the trigonometric form. But we see that negative frequency components are introduced and, since these are physically meaningless, the question must be asked, can the results of the complex Fourier series be interpreted in a way that is physically acceptable, i.e. can the two-sided spectrum be transformed into a one-sided representation?

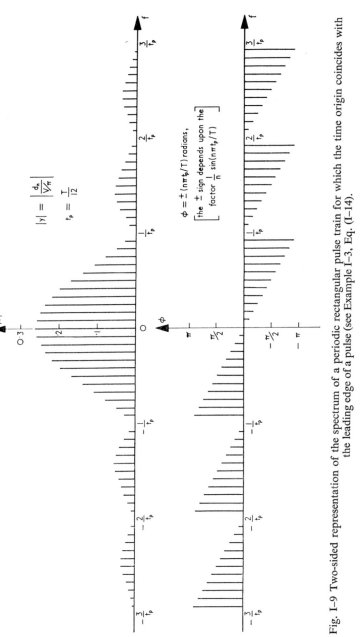

Fig. I-9 Two-sided representation of the spectrum of a periodic rectangular pulse train for which the time origin coincides with the leading edge of a pulse (see Example I-3, Eq. (I-14).

From Eq. (I–8),

$$f(t) = \sum_{n=-\infty}^{\infty} d_n \exp(jn\omega t)$$

$$f(t) = d_0 + \sum_{n=1}^{\infty} d_n \exp(jn\omega t) + \sum_{n=1}^{\infty} d_{-n} \exp(-jn\omega t)$$

$$= d_0 + \sum_{n=1}^{\infty} d_n \exp(jn\omega t) + \sum_{n=1}^{\infty} d_n^* \exp(-jn\omega t)$$

Using Eq. (I–9a) and substituting for d_n,

$$f(t) = d_0 + \sum_{n=1}^{\infty} \tfrac{1}{2} c_n \exp(-j\phi_n)\exp(jn\omega t) + \sum_{n=1}^{\infty} \tfrac{1}{2} c_n \exp(j\phi_n)\exp(-jn\omega t)$$

$$= d_0 + \sum_{n=1}^{\infty} \tfrac{1}{2} c_n (2\cos\phi_n \cos n\omega t + 2\sin\phi_n \sin n\omega t)$$

$$= d_0 + \sum_{n=1}^{\infty} 2\,\mathrm{Re}[d_n \exp(jn\omega t)] \qquad (I\text{–}15)$$

The result can be checked by considering the previous example of a rectangular pulse train. When the time origin is taken at the centre of a pulse the coefficient d_n is given by Eq. (I–11). If this is substituted into Eq. (I–15) the result is identical to that of the trigonometric analysis (see Eq. (I–7)). A more convincing check can be made when d_n is complex, as shown by the next example.

Example I–4. Consider the rectangular pulse train with the time origin coinciding with the leading edge of a pulse. The coefficient d_n is given by Eq. (I–13),

$$d_n = \frac{V}{n\pi} \sin(n\pi t_p/T)\exp(-jn\pi t_p/T)$$

Substituting into Eq. (I–15) gives,

$$f(t) = d_0 + \sum_{n=1}^{\infty} 2\mathrm{Re}\left[\frac{V}{n\pi} \sin(n\pi t_p/T)\exp(-jn\pi t_p/T)\exp(jn\omega t)\right]$$

$$= d_0 + \sum_{n=1}^{\infty} \frac{V}{n\pi} \sin(2\pi n t_p/T)\cos n\omega t + \sum_{n=1}^{\infty} \frac{V}{n\pi}\left[1 - \cos(2\pi n t_p/T)\right]\sin n\omega t$$

$$(I\text{–}16)$$

To verify that this one-sided spectral representation is correct, let us evaluate the Fourier spectrum from the trigonometric form of analysis. Referring to Fig. I–8, the period BB is defined as

$$f(t) = V, \quad 0 < t < t_p$$

$$f(t) = 0, \qquad t > t_p$$

Therefore,

$$a_n = \frac{2}{T} \int_0^{t_p} V \cos n\omega t \, dt = \frac{V}{n\pi} \sin(2\pi n t_p / T)$$

and

$$b_n = \frac{2}{T} \int_0^{t_p} V \sin n\omega t \, dt = \frac{V}{n\pi}\left[1 - \cos(2\pi n t_p / T)\right]$$

Therefore, using Eq. (I–2), the general expression for the periodic waveform is

$$f(t) = \frac{V t_p}{T} + \sum_{n=1}^{\infty} \frac{V}{n\pi} \sin(2\pi n t_p / T) \cos n\omega t + \sum_{n=1}^{\infty} \frac{V}{n\pi}\left[1 - \cos(2\pi n t_p / T)\right] \sin n\omega t$$

and this expression is identical to Eq. (I–16).

I–4 Frequency Spectrum of a Single Pulse—Fourier Integral

So far we have considered a periodic waveform that commenced at $t = -\infty$ and continues to $t = +\infty$, a physically unrealizable condition, but, nevertheless, one which was necessary in order to allow the relatively simple application of the Fourier series. A more meaningful condition can be obtained by considering the period T to become very large. Then, for a series of rectangular pulses, as $T \to \infty$, we are concerned with a single pulse in the time domain. Applying this condition to a periodic function means that the frequency spacing between the spectral components becomes smaller as T gets larger and at the limit, $T \to \infty$, the line spectrum becomes a continuous spectrum. If we take Eq. (I–10), which defines the line spectrum of a periodic function, as T becomes large the frequency spacing of the components is $\delta f = 1/T$ and

$$d_n = \delta f \int_{-T/2}^{T/2} f(t) \exp(-j\omega t) \, dt \qquad (I–17)$$

where ω is used in place of $n\delta\omega$ in the exponential function. Consequently as $T \to \infty$

$$\lim_{T \to \infty} \frac{d_n}{1/T} = \int_{-\infty}^{\infty} f(t) \exp(-j\omega t) \, dt$$

$$[I–4]$$

The validity of the mathematical operation will not be considered here; a full treatment can be found elsewhere.[1] For our purposes the integral in the above equation must be finite for any real value of ω. In short, a condition that can be satisfied if $\int_{-\infty}^{\infty} f(t)\,dt$ is absolutely convergent.

It is known as the Fourier Integral and is given the symbol $F(j\omega)$, denoting that in general it is complex and a function of frequency.

$$F(j\omega) = \int_{-\infty}^{\infty} f(t)\exp(-j\omega t)\,dt \qquad (\text{I–18})$$

It can be seen that $F(j\omega)$ has the dimensions of volt-seconds or volts/Hz, when $f(t)$ defines a voltage-time variation, and is therefore an amplitude spectral density function. (For the purposes of illustration throughout the appendix, $f(t)$ will be considered to have the dimensions of volts.) Before we attempt to explain this function in terms of something which is physically meaningful, it must be borne in mind that the continuous spectrum is the line spectrum with infinitesimally small spacing between the frequency components. But it is no longer possible to consider a discrete sinusoid at some specific frequency, since its amplitude would be zero (see Eq. (I–17))—as $\delta f \to 0$, then $d_n \to 0$. If it is not possible to distinguish discrete sinusoids in the frequency domain, the question may be asked, 'does $F(j\omega)$ have a physically recognisable identity?'

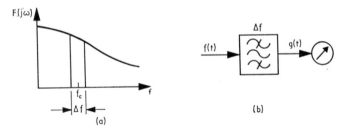

Fig. I–10. Typical amplitude-density spectrum of a non-periodic signal—the shaded area is that part of the spectrum selected by the band-pass filter.

In order to answer this question, let us consider a plot of $F(j\omega)$ against frequency. The area under the curve will have the dimensions of volts and for a small but finite bandwidth Δf at frequency f_c (see Fig. I–10 (a)) the area is approximately $F(j\omega)\Delta f$. This could be realised in practice by applying a time varying non-periodic (by physical definition) function, whose Fourier transform is $F(j\omega)$, to a narrow-bandpass

1. *Laplace and Fourier Transforms for Electrical Engineers*, E. J. Craig (Holt, Rinehart and Winston, 1964).

filter (Fig. I–10 (*b*)). The output from the filter will depend upon $F(j\omega)\Delta f$, but it will also be a function of time. Consider, for example, that $f(t)$ defines a single rectangular pulse. The output will vary in sympathy with the input, rising from zero to some maximum value and then decreasing back to zero. That is to say, we could only attach some physically realizable significance to the time variations of the output $g(t)$, and that although this depends upon $F(j\omega)\Delta f$, the latter cannot by itself be given a meaningful identity. We conclude, therefore, that the amplitude spectral density function $F(j\omega)$ is the result of a mathematical operation with apparently no physically tangible meaning. As an operator, however, it has some extremely useful applications, two of which will be considered in detail, namely,

(i) the evaluation of the output response of a linear network when the input is known, and

(ii) the spectral distribution of power or energy for a given time-varying function.

Example I–5. To find the Fourier transform $F(j\omega)$ of a single rectangular pulse of amplitude V and duration t_p.

Consider the time origin to coincide with the leading edge of the pulse. We shall see that this is not the preferred position and that the simplest form of $F(j\omega)$ is obtained when there is symmetry about the time origin.

$$f(t) = V, \quad 0 < t < t_\mathrm{p}$$

$$f(t) = 0, \quad \text{for all other } t$$

Therefore
$$F(j\omega) = \int_0^{t_\mathrm{p}} V \exp(-j\omega t)dt$$

$$= \frac{2V}{\omega} \exp(-j\omega t_\mathrm{p}/2)\sin(\omega t_\mathrm{p}/2)$$

$$= Vt_\mathrm{p} \frac{\sin x}{x} \exp(-jx) \text{ volt-seconds}$$

where $x = \omega t_\mathrm{p}/2$. It is convenient to express $F(j\omega)$ in the form shown by the last line, because the $\sin x/x$ factor is well known and lends itself to graphical illustration. In this instance a plot of $F(j\omega)$ against frequency will not be performed, since it has been shown that the transform has no physically meaningful identity. However, it does have a useful role in the analysis of linear networks and therefore it is desirable to obtain the simplest form of expressing $F(j\omega)$ for a given time-varying function. This may be achieved by positioning the time origin at the centre of the pulse; compare this with the treatment of the periodic function in section I–2.

$$f(t) = V, \quad -t_\mathrm{p}/2 < t < t_\mathrm{p}/2$$

$$f(t) = 0, \quad \text{for all other } t.$$

$$[\text{I–4}]$$

Therefore, $F(j\omega) = \displaystyle\int_{-t_p/2}^{t_p/2} V\exp(-j\omega t)dt$

$$= Vt_p \frac{\sin(\omega t_p/2)}{(\omega t_p/2)} \text{ volt-seconds.} \qquad (I\text{-}19)$$

Tutorial Exercise I–3. Obtain the Fourier transform of a single triangular pulse of peak amplitude V and duration t_p.

$$Ans: F(j\omega) = 2Vt_p\left[\frac{\sin(\omega t_p/4)}{(\omega t_p/2)}\right]^2$$

Example I–5 shows that when the pulse is displaced in time by $t_p/2$ the Fourier transforms differ by a factor $\exp(-j\omega t_p/2)$. The time displacement of a function has many applications and it is worth while to develop the general case. The Fourier transform of $f(t)$ is

$$\mathfrak{F}\,[f(t)] = F(j\omega) = \int_{-\infty}^{\infty} f(t)\exp(-j\omega t)\,dt$$

and if the function is delayed by a time T_d then $f(t)$ becomes $f(t-T_d)$ and its Fourier transform is

$$\mathfrak{F}\,[f(t-T_d)] = \int_{-\infty}^{\infty} f(t-T_d)\exp(-j\omega t)\,dt$$

By letting $x = t-T_d$,

$$\mathfrak{F}\,[f(t-T_d)] = \int_{-\infty}^{\infty} f(x)\exp(-j\omega x)\exp(-j\omega T_d)\,dx$$

$$= \left\{\int_{-\infty}^{\infty} f(x)\exp(-j\omega x)\,dx\right\}\exp(-j\omega T_d)$$

$$= F(j\omega)\exp(-j\omega T_d)$$

Therefore, the time displacement of a function leaves its amplitude spectrum unchanged, but modifies the so-called phase spectrum by a factor $\exp(-j\omega T_d)$ (see also Example I–3, page 215).

A particularly useful application of the above example is obtained when the rectangular pulse has an infinitesimally small duration. This condition defines the impulse function, given by $\displaystyle\lim_{t_p\to 0} f(t)$, but before the Fourier transform can be evaluated the amplitude and the product of amplitude-time duration of the pulse require further definition—there are two possibilities,

(i) a physically unrealizable but mathematically convenient concept assumes that $V\to\infty$ and the product $Vt_p\to 1$. This defines the unit impulse function $\delta(t)$, or Dirac delta function, as it is also known, such that,

$$\lim_{\substack{t_p \to 0 \\ V \to \infty}} f(t) = \delta(t)$$

where $\delta(t) = 0$ for all values of t except $t = 0$ and $\displaystyle\int_{-\infty}^{\infty} \delta(t)\, dt = 1$.

The Fourier transform of the unit impulse function is obtained from Eq. (I–19),

$$F_\delta(j\omega) = \lim_{\substack{t_p \to 0 \\ V \to \infty \\ Vt_p = 1}} \left\{ Vt_p \frac{\sin(\omega t_p/2)}{\omega t_p/2} \right\}$$

$$= \lim_{Vt_p = 1} Vt_p = 1$$

(ii) a physically acceptable condition regarding the pulse amplitude V and pulse 'area' Vt_p assumes that V remains finite and that Vt_p is equal to a constant but not necessarily unity, as in the previous definition. For mathematical purposes, the pulse duration is still considered to be infinitesimally small, and in practice a very good approximation to this condition can be realised so long as the pulse duration is exceedingly small compared with the time constant or constants of the network to which it is applied. In this case

$$F(j\omega) = \lim_{\substack{t_p \to 0 \\ Vt_p = \text{const.}}} \left\{ Vt_p \frac{\sin(\omega t_p/2)}{\omega t_p/2} \right\}$$

$$= constant.$$

We see, therefore, that the impulse function has a uniform spectral density, or that the frequency spectrum is flat, for either $Vt_p = 1$ or $Vt_p = constant$. Application of this function is made in subsequent sections of the appendix.

I–5 Response of Linear Channels

Before we attempt to use the preceeding analysis it is firstly necessary to define the network or channel under consideration. The term linear is used to define a circuit when the following conditions apply:

(i) the output voltage and current are directly proportional to the input voltage and current, and

(ii) if two signals are applied simultaneously the output signal is the

[I–5]

sum of the two outputs obtained had the inputs been applied individually —the Superposition Theorem.

These conditions may be re-stated as follows. If a sinusoidal signal or a number of sinusoidal components representing a signal are applied to a linear network, the output is made up of only these components either suitably amplified or attenuated. If sinusoidal components at frequencies other than those applied to the input appear at the output, then the circuit is said to have a non-linear characteristic.

In practice all networks (circuits or channels) exhibit some degree of non-linearity, but often this may be ignored when the characteristics are predominantly linear. For example, a small signal transistor amplifier is regarded as a linear amplifier when the distortion (or non-linearity) is small. On the other hand, a diode used as a modulator must exhibit a significant non-linear characteristic in order to perform the required function, and hence it would violate the conditions postulated for a linear circuit.

The characteristics of the linear channel may be specified as the ratio V_{out}/V_{in} when the input voltage V_{in} is a single sine wave. This ratio over the frequency range of interest gives information regarding the amplitude-frequency response $A(\omega)$ of the network and also its phase-frequency response $\phi(\omega)$, i.e.,

$$\frac{V_{out}}{V_{in}} = A(\omega)\exp j\phi(\omega)$$

The ratio is known as the transfer function of the network and in order to simplify the appearance of the mathematics it shall be referred to as $H(j\omega)$ denoting that, in general, it is complex and a function of frequency.

Example I–6. Evaluation of the transfer functions of three simple networks.

(i) Two perfect resistors forming a potential divider.

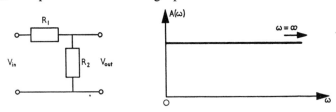

Fig. I–11 Amplitude characteristic of a resistive divider.

The transfer function is independent of frequency, also the output voltage is in phase with the input voltage.

Therefore $\quad\quad\quad\quad H(j\omega) = A(\omega) = \dfrac{R_2}{R_1+R_2}$

(ii) A resistor-capacitor divider.

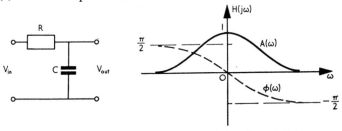

Fig. I–12 Amplitude and phase characteristics of a *RC* divider.

$$H(j\omega) = \frac{1/j\omega C}{R+1/j\omega C}$$

$$= \frac{1}{1+j\omega CR}$$

It is usually convenient to leave the transfer function in this form, but if required in polar form then

$$H(j\omega) = \frac{1}{\sqrt{1+\omega^2 C^2 R^2}} \exp[-j\tan^{-1}(\omega CR)]$$

Note that Fig. I–12 shows $H(j\omega)$ for both positive and negative frequencies. This is necessary if the transfer function is to be used in a Fourier analysis involving the complex form of representation.

(iii) An idealised low-pass filter with cut-off frequency f_0 and constant time delay T_d. Thus the amplitude characteristic is defined by

$$A(\omega) = A, \quad |f| < |f_0|$$
$$A(\omega) = 0 \quad |f| > |f_0|$$

The modulus of the frequency is considered so that positive and negative frequencies are defined.

The phase characteristic will be linear, due to the constant time delay T_d, and is given by

$$\phi(\omega) = -\omega T_d$$

where the negative sign indicates that the output is lagging with respect to the input. It is this linear phase relationship that cannot be realised in practice— further explanation is given in Example I–7.

Therefore, for the idealised low-pass filter,

$$H(j\omega) = A\exp(-j\omega T_d), \quad |f| < |f_0|$$

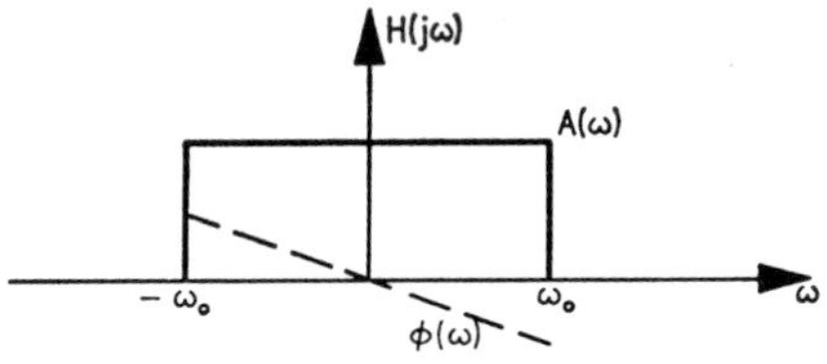

Fig. I–13 Amplitude and phase characteristics of an ideal low-pass filter.

[I–5]

In the example above it is stated that $H(j\omega)$ must be defined for positive and negative frequencies if it is to be used in a Fourier analysis involving the complex form of notation. This is achieved by simply substituting ω for $-\omega$ for the negative frequencies. Although this may appear to be a reasonable and, in fact, intuitive procedure, it is essential to check that for a real input $f(t)$, applied to a network whose transfer function is defined in this way, the output response $g(t)$ is also a real function.

Consider, for example, an input $f(t) = \cos\omega_c t$. Since this is a periodic function, the Fourier coefficient d_n may be evaluated. But the spectrum will only consist of two lines, at $+\omega_c$ and $-\omega_c$; therefore $d_n = \frac{1}{2}$ and

$$f(t) = \tfrac{1}{2}\exp(j\omega_c t) + \tfrac{1}{2}\exp(-j\omega_c t)$$

(this result could be obtained directly from the exponential form of expressing the trigonometric function $\cos \omega t$, but it is a constructive exercise and a useful check to show that the result can be obtained from the formal Fourier approach).

The output $g(t)$ from a network with transfer function $H(j\omega)$, is

$$g(t) = \tfrac{1}{2}H(j\omega_c)\exp(j\omega_c t) + \tfrac{1}{2}H(-j\omega_c)\exp(-j\omega_c t)$$

and since $g(t)$ must be a real function,

$$H^*(j\omega_c) = H(-j\omega_c)$$

This may be shown to be a necessary condition for any real input $f(t)$ if the output is also to be a real function. Therefore, in general, the transfer function at negative frequencies must equal the complex conjugate of the function representing the positive frequency response, i.e.

$$H^*(j\omega) = H(-j\omega) \tag{I-20}$$

When a non-periodic function with Fourier transform $F(j\omega)$ is applied to the input of a linear network, the amplitude-frequency spectrum of the output is given by

$$G(j\omega) = F(j\omega)H(j\omega)$$

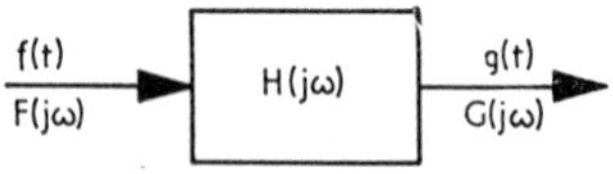

Fig. I–14 Linear network with transfer function $H(j\omega)$.

We may now ask the question, 'Can the time-varying output $g(t)$ be obtained from $G(j\omega)$?' Or, in general, 'Can the inverse operation of converting $F(j\omega)$ back to its equivalent time-varying function $f(t)$ be performed?' To answer this, consider the periodic function $f(t)$ defined by

$$f(t) = \sum_{n=-\infty}^{\infty} d_n \exp(jn\omega t)$$

$$= \sum_{n=-\infty}^{\infty} \frac{1}{T}\left\{\int_{-T/2}^{T/2} f(t)\exp(-jn\omega t)dt\right\} \exp(jn\omega t)$$

When $T \to \infty$

$$f(t) = \int_{-\infty}^{\infty} df\left\{\int_{-\infty}^{\infty} f(t)\exp(-j\omega t)dt\right\} \exp(j\omega t)$$

Therefore $f(t) = \int_{-\infty}^{\infty} F(j\omega)\exp(j\omega t)df$ \qquad (I–21)

This is known as the inverse Fourier transform. Therefore the answer to the question is 'yes', and the time-varying output $g(t)$ of the linear network is given by

$$g(t) = \int_{-\infty}^{\infty} G(j\omega)\exp(j\omega t)df$$

$$= \int_{-\infty}^{\infty} F(j\omega)H(j\omega)\exp(j\omega t)df$$

Unfortunately, this integral is often difficult to evaluate; take, for example, the relatively simple case of a rectangular pulse applied to a *RC* filter.

$$F(j\omega) = Vt_p \frac{\sin(\omega t_p/2)}{\omega t_p/2}$$

for the time origin at the centre of the pulse, and

$$H(j\omega) = \frac{1}{1+j\omega CR}$$

Therefore, $g(t) = \int_{-\infty}^{\infty} Vt_p \dfrac{\sin(\omega t_p/2)}{\omega t_p/2} \dfrac{1}{1+j\omega CR} \exp(j\omega t)\, df$

It is not recommended to evaluate this integral, since there are simpler methods of determining $g(t)$, namely the use of the Convolution Integral (section I–8), or by the use of tables of Laplace transforms with the operator $j\omega$ substituted in place of the operator s. The latter method is

[I–5]

not dealt with in this book, but the references are recommended for a comparison of Fourier and Laplace methods.[2,3] Briefly, Fourier transforms are usually preferred for communications analyses, since the duality between time and frequency is established—the concept of bandwidth and how it is utlised is of paramount importance. Laplace methods lend themselves more readily to transient analysis.

Example I–7. Consider a rectangular pulse of amplitude V and duration t_p applied to an idealised low-pass filter. We wish to obtain the Fourier transform $G(j\omega)$ of the output and from this evaluate the output response $g(t)$.

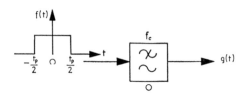

Fig. I–15 Rectangular pulse applied to an ideal low-pass filter.

For the pulse, $\quad F(j\omega) = Vt_\mathrm{p}\dfrac{\sin(\omega t_\mathrm{p}/2)}{\omega t_\mathrm{p}/2}$

and for the filter, $\quad H(j\omega) = A\exp(-j\omega T_\mathrm{d}), \quad |f| < |f_0|$

The Fourier transform of the output is

$$G(j\omega) = F(j\omega)\,H(j\omega) = AVt_\mathrm{p}\frac{\sin(\omega t_\mathrm{p}/2)}{\omega t_\mathrm{p}/2}\exp(-j\omega T_\mathrm{d})$$

and using the inverse Fourier transform, Eq. (I–21),

$$g(t) = AVt_\mathrm{p}\int_{-f_0}^{f_0}\frac{\sin(\omega t_\mathrm{p}/2)}{\omega t_\mathrm{p}/2}\exp[j\omega(t-T_\mathrm{d})]df$$

To evaluate this integral, firstly use the relationship

$$\exp(j\theta) = \cos\theta + j\sin\theta$$

then $\quad g(t) \quad = AVt_\mathrm{p}\displaystyle\int_{-f_0}^{f_0}\frac{\sin(\omega t_\mathrm{p}/2)}{\omega t_\mathrm{p}/2}\cos[\omega(t-T_\mathrm{d})]df$

$$+ jAVt_\mathrm{p}\int_{-f_0}^{f_0}\frac{\sin(\omega t_\mathrm{p}/2)}{\omega t_\mathrm{p}/2}\sin[\omega(t-T_\mathrm{d})]df$$

The integrand of the second integral is an odd function and is therefore equal

2. *Laplace Transforms for Electronic Engineers*, J. G. Holbrook (Pergamon, 1959).
3. *Handbook of Laplace Transformation*, F. E. Nixon (Prentice-Hall, 1965).

to zero between the limits f_0 and $-f_0$. The integrand of the first integral is an even function, therefore

$$g(t) = 2AVt_p \int_0^{f_0} \frac{\sin(\omega t_p/2)}{\omega t_p/2} \cos [\omega(t-T_d)]df$$

$$= AVt_p \int_0^{f_0} \left\{ \frac{\sin\omega(t-T_d+t_p/2)}{\omega t_p/2} - \frac{\sin\omega(t-T_d-t_p/2)}{\omega t_p/2} \right\} df$$

Finally, by changing the variables, with $x = 2\pi f(t-T_d+t_p/2)$ in the first integral and $y = 2\pi f(t-T_d-t_p/2)$ in the second,

$$g(t) = \frac{AV}{\pi} \int_0^{\omega_0(t-T_d+t_p/2)} \frac{\sin x}{x} dx - \frac{AV}{\pi} \int_0^{\omega_0(t-T_d-t_p/2)} \frac{\sin y}{y} dy$$

The sin x/x integral cannot be evaluated in closed form, but tables are available,[4, 5]

$$\text{Si } x_1 = \int_0^{x_1} \frac{\sin x}{x} dx$$

The output response is usually written as

$$g(t) = \frac{AV}{\pi} \{ \text{Si}[\omega_0(t-T_d+t_p/2)] - \text{Si}[\omega_0(t-T_d-t_p/2)] \}$$

Fig. I–16 shows a plot of this result for three values of the filter bandwidth in terms of the pulse duration. The outputs are symmetrical about a time delay T_d, but their shapes differ considerably.

(i) When $f_0 \ll 1/t_p$, the output is a 'smeared' version of the input, i.e. gross distortion has occurred.

(ii) For the condition $f_0 = 1/t_p$, the pulse is recognisable, but the rise and fall times are significant compared with the pulse duration.

(iii) When $f_0 \gg 1/t_p$, the output has approximately the same pulse width as the input, but there are several marked differences. Firstly, the rise and fall do not occur in zero time and secondly, there is considerable overshoot or ringing at the leading and trailing edges, a characteristic of all filters with sharp cut-off amplitude responses. The damped oscillation forming the tail of the pulse is in no way surprising, but the oscillation preceding the pulse means that the output appears before the input has been applied! This physically impossible precursor is due to the idealised linear phase relationship that has been assumed for the filter. In practice this can only be approached by using a large number of components in distributed form rather than a few in lumped form. In order to obtain a true linear phase relationship, an infinite number of circuit elements would be required, resulting in an output delay T_d which becomes infinite. The general treatment of transfer functions that are physically realizable is beyond the scope of this book; however a full treatment can be found elsewhere.[6, 7]

4. *Mathematical Tables*, Vol. 1, Brit. Ass. Adv. Sc. (London, 1931).
5. *Tables of Si(x) and Ci(x)*, K. Tani (Meguro, Tokyo, 1931).
6. *Network Analysis and Synthesis*, Chapter 10, F. F. Kuo (Wiley, 1966).
7. *Introduction to Modern Network Synthesis*, M. E. Van Valkenburg (Wiley, 1960).

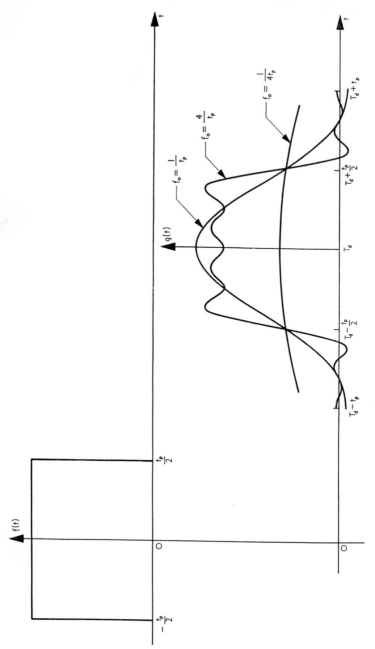

Fig. I-16 Response of an ideal low-pass filter to a rectangular pulse.

It can be seen from the above example that the output is distorted for the three conditions of filter cut-off frequency/pulse duration. We are therefore prompted to determine the characteristics of a linear circuit that would have an output response $g(t)$ of identical shape to the input, i.e. one that would allow distortionless transmission. Such a condition means that

$$g(t) = Kf(t-T_d)$$

where the constant K allows an amplification or attenuation of the input and T_d represents the time delay—but the time response is unaltered. The Fourier transform of the output is

$$G(j\omega) = \int_{-\infty}^{\infty} Kf(t-T_d)\exp(-j\omega t)dt$$

But $\int_{-\infty}^{\infty} f(t-T_d)\exp(-j\omega t)dt$ is the Fourier transform of $f(t)$ with a shift of the time origin to the position $t = T_d$ (see page 222).

Therefore $\qquad \int_{-\infty}^{\infty} f(t-T_d)\exp(-j\omega t)dt = F(j\omega)\exp(-j\omega T_d)$

hence $\qquad G(j\omega) = KF(j\omega)\exp(-j\omega T_d)$

Also $G(j\omega)$ and $F(j\omega)$ are related by the network transfer function $H(j\omega)$,

$$G(j\omega) = H(j\omega)F(j\omega)$$

A comparison of the two equations shows that $H(j\omega)$ must be equal to $K\exp(-j\omega T_d)$ to satisfy the condition of distortionless transmission. This means that the linear network must have a constant amplitude characteristic over the frequency spectrum of the input signal and a phase-frequency characteristic that is linear over the same range of frequencies.

For the rectangular pulse, distortionless transmission has no meaning, since the channel would have to be capable of passing all the frequencies of the spectrum. In fact, a pulse with zero rise and fall times cannot be generated; consequently a discussion of the transmission requirements is indeed hypothetical. In practice, all pulses are to some extent band-limited, although it is often convenient to consider idealised pulse shapes for ease of mathematical handling. *It cannot be stressed too strongly that it is quite in order to make such approximations so long as the idealisations are kept in perspective and the results used simply to indicate what is likely to occur in reality.*

[I–5]

The recurring theme in the design of pulse modulation systems is one of bandwidth economy against the conflicting requirement of minimum pulse distortion. Example I–7 serves as an illustration. An intentional departure from the distortionless transmission requirement has been made by placing a cut-off frequency f_0 to the amplitude response of the network. As bandwidth is conserved by making f_0 smaller, the distortion of the output increases. If the phase-frequency characteristic is assumed linear, a requirement for distortionless transmission but one which can never be realised in practice, the output response is symmetrical about $t = T_d$, regardless of the degree of amplitude distortion. When the phase response is non-linear, a necessary condition for the filter to be physically realizable, the output is unsymmetrical and the precursor condition is avoided. The two types of distortion are shown in Fig. I–17.[8]

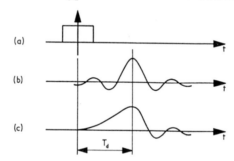

Fig. I–17 Network amplitude and phase distortion.
(*a*) Input pulse,
(*b*) output response, amplitude distortion, linear phase response, and
(*c*) output response, non-symmetrical phase response.

I–6 Impulse Response of a Linear Channel—Equivalence of Low-pass and Band-pass Filters

The analysis of the previous section has useful applications when the input is an impulse function with Fourier transform $F(j\omega) = constant = k$, say (see section I–4).

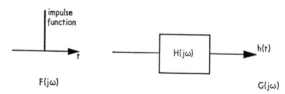

Fig. I–18 Impulse function applied to a linear network.

8. *Pulses and Transients in Communication Circuits*, C. Cherry, p. 145 (Chapman and Hall, 1949).

The Fourier transform of the output is

$$G(j\omega) = F(j\omega)H(j\omega)$$
$$= kH(j\omega)$$

[or $G(j\omega) = H(j\omega)$ when the input is a unit impulse function].

The output response, which can be obtained from the inverse Fourier transform of $G(j\omega)$, is given the symbol $h(t)$, in place of $g(t)$, to denote that the input is an impulse. Therefore

$$h(t) = \int_{-\infty}^{\infty} kH(j\omega)\exp(j\omega t)df$$

Example I–8. The impulse response of the ideal low-pass filter.

For the delta impulse, $\quad F_\delta(j\omega) = 1$ $\Big\}$ (see Fig. I–19),
For the pseudo-impulse, $\quad F(j\omega) = k$

and for the filter, $\qquad H(j\omega) = A\exp(-j\omega T_d), \quad |f| < |f_0|$

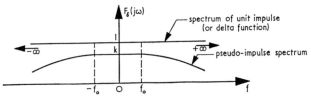

Fig. I–19 Frequency spectra of impulse functions.

Therefore the impulse response is

$$h(t) = \int_{-f_0}^{f_0} kA\exp[j\omega(t-T_d)]df = 2kAf_0\frac{\sin[\omega_0(t-T_d)]}{\omega_0(t-T_d)} \qquad (I–22)$$

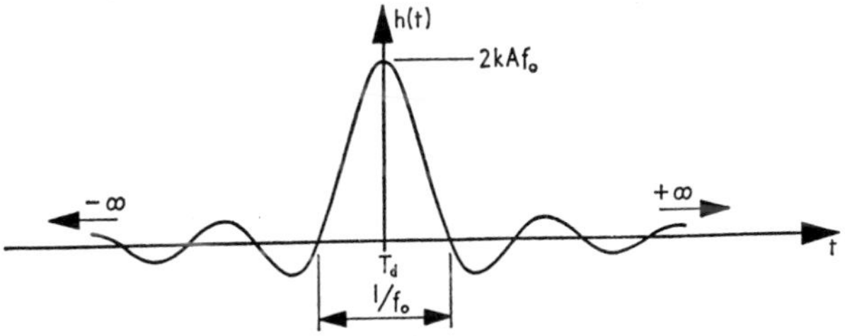

Fig. I–20 Impulse response of an ideal low-pass filter.

Tutorial Exercise I–4. Check that Eq. (I–22) is dimensionally correct.

Having considered the response of the ideal low-pass filter, a reasonable progression would be to evaluate next the response of a band-pass filter.

[I–6]

In communication systems this type of filter is used to select a wanted signal from a frequency spectrum of many signals and accompanying wideband noise. The required signal is invariably a modulated carrier at the centre frequency of the band-pass filter and therefore it is this type of input that should be used in the succeeding analysis.

Although this section is concerned with the impulse responses of networks, it is desirable to make a short digression to consider a specific example of a modulated signal.

Example I–9. The find the Fourier transform of a carrier modulated by a single pulse of duration t_p, and defined as $V\sin\omega_c t$, see Fig. I–21, or $V\cos\omega_c t$ during the interval $-t_p/2 < t < t_p/2$.

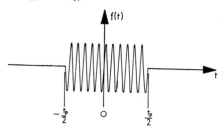

Fig. I–21 Carrier modulated by a rectangular pulse of duration t_p.

The Fourier transform is

$$\mathfrak{F}[V\cos\omega_c t] = \int_{-t_p/2}^{t_p/2} V\cos\omega_c t\exp(-j\omega t)dt$$

$$= \frac{V}{2}\int_{-t_p/2}^{t_p/2}\exp[-j(\omega-\omega_c)t]dt + \frac{V}{2}\int_{-t_p/2}^{t_p/2}\exp[-j(\omega+\omega_c)t]dt$$

$$= V\frac{\sin[(\omega-\omega_c)t_p/2]}{\omega-\omega_c} + V\frac{\sin[(\omega+\omega_c)t_p/2]}{\omega+\omega_c}$$

$$= \frac{Vt_p}{2}\frac{\sin[(\omega-\omega_c)t_p/2]}{(\omega-\omega_c)t_p/2} + \frac{Vt_p}{2}\frac{\sin[(\omega+\omega_c)t_p/2]}{(\omega+\omega_c)t_p/2}$$

But the above equation defines a two-sided frequency representation, that is, ω can have positive and negative values. Consider the first term on the right-hand side: when $\omega = +\omega_c$ the term is a maximum and is equal to $Vt_p/2$. The same term with $\omega = -\omega_c$ will be very much smaller and therefore its contribution to the negative frequency spectrum can be ignored. Similarly, the second term on the right-hand side of the equation has its maximum value of $Vt_p/2$ when $\omega = -\omega_c$ and is very small when $\omega = +\omega_c$. Therefore,

$$\mathfrak{F}[V\cos\omega_c t] \approx \frac{Vt_p}{2}\frac{\sin[(\omega-\omega_c)t_p/2]}{(\omega-\omega_c)t_p/2}\bigg|_{\substack{\omega \\ \text{positive}}} + \frac{Vt_p}{2}\frac{\sin[(\omega+\omega_c)t_p/2]}{(\omega+\omega_c)t_p/2}\bigg|_{\substack{\omega \\ \text{negative}}} \quad \text{(I–23)}$$

The approximation only holds good if $1/t_p \ll f_c$, i.e. the rectangular modulating pulse is supported by many cycles of the carrier. It is interesting to note that the Fourier transform of the modulated signal is of identical form, but half the magnitude of the Fourier transform of the modulating (baseband) signal, in this case a single pulse (cf. Eq. (I–19)), and that it is centred on $+f_c$ and $-f_c$ in place of zero frequency.

This somewhat awkward solution, in which the positive and negative frequency spectra of the modulated signal overlap, is due to the modulating pulse having an infinite spectrum. In practice, this would never arise, since the modulating (baseband) signal is always band-limited and the carrier frequency is numerically greater than the baseband. Therefore, let us establish the general result for the physically realizable condition of a carrier modulated by a band-limited signal $m(t)$ and whose Fourier transform is defined as

$$\mathfrak{F}[m(t)] = M(j\omega)$$

then $$\mathfrak{F}[m(t)\exp(i\omega_c t)] = \int_{-\infty}^{\infty} m(t)\exp[-j(\omega-\omega_c)t]dt$$

$$= M[j(\omega-\omega_c)] \qquad (I\text{--}24)$$

where the bracket $(\omega-\omega_c)$ denotes that it is the Fourier transform of $m(t)$ shifted in frequency to ω_c (cf. the time displacement of a function, page 222). Therefore, the Fourier transform of the modulated carrier is

$$\mathfrak{F}[m(t)\cos\omega_c t] = \mathfrak{F}[\tfrac{1}{2}m(t)\exp(j\omega_c t)+\tfrac{1}{2}m(t)\exp(-j\omega_c t)]$$

$$= \tfrac{1}{2}M[j(\omega-\omega_c)]+\tfrac{1}{2}M[j(\omega+\omega_c)] \qquad (I\text{--}25)$$

The amplitude spectra of a baseband signal and corresponding modulated signal are shown in Fig. I–22.

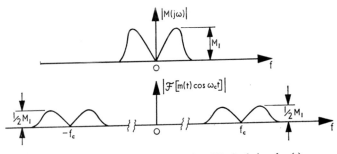

Fig. I–22   (*a*) Amplitude spectrum of a band-limited signal $m(t)$.
(*b*) Amplitude spectrum of a modulated carrier, $m(t)\cos\omega_c t$.

[I–6]

Previously, for the analysis of the ideal low-pass filter, its response was evaluated for an input consisting of a single rectangular pulse (see page 228). We are about to consider a band-pass filter and, with the above comment in mind, it would be reasonable and indeed logical to evaluate its response to an input consisting of a carrier at the centre frequency of the filter and modulated by a rectangular pulse. Let the band-pass filter be defined as follows:

$$H(j\omega) = A\exp(-j\omega T_d) \begin{cases} f_c - \Delta/2 < f < f_c + \Delta/2 \\ -f_c - \Delta/2 < f < -f_c + \Delta/2 \end{cases}$$

and $\qquad H(j\omega) = 0$ for all other frequencies.

Note the condition that the output response should be a real function is satisfied, i.e. $H^*(j\omega) = H(-j\omega)$.

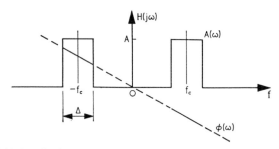

Fig. I–23 Amplitude and phase characteristics of an ideal band-pass filter.

The Fourier transform of a carrier modulated by a rectangular pulse is given by Eq. (I–23). Therefore, the output response of the band-pass filter is

$$g(t) \approx \int_{f_c - \Delta/2}^{f_c + \Delta/2} \frac{AVt_p}{2} \frac{\sin[(\omega - \omega_c)t_p/2]}{(\omega - \omega_c)t_p/2} \exp[j\omega(t - T_d)]df$$

$$+ \int_{-f_c - \Delta/2}^{-f_c + \Delta/2} \frac{AVt_p}{2} \frac{\sin[(\omega + \omega_c)t_p/2]}{(\omega + \omega_c)t_p/2} \exp[j\omega(t - T_d)]df$$

Regrettably its evaluation is heavy work and will not be attempted here. The integral could be simplified by making the Fourier transform of the modulated signal a constant over the passband of the filter in place of using the expression given by Eq. (I–23). The modulating signal is, therefore, a pseudo-impulse function,* and furthermore, we shall consider it to be band-limited so that the analysis leading to Eq. (I–25)

---

* The pseudo-impulse function is the physically meaningful interpretation of the delta (unit) impulse function (see page 233).

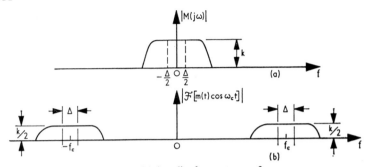

Fig. I–24 Amplitude spectrum of
(a) a pseudo-impulse (band-limited) modulating signal, and
(b) the amplitude modulated carrier.

may be applied (see Fig. I–24, and cf. Fig. I–19). From Eq. (I–25), the Fourier transform of the modulated carrier may be defined over the passband of the filter as

$$\frac{1}{2}M[j(\omega-\omega_c)]\Big|_{(f_0-\Delta/2)<f<(f_0+\Delta/2)} = \frac{1}{2}M[j(\omega+\omega_c)]\Big|_{(-f_0-\Delta/2)<f<(-f_0+\Delta/2)} = \frac{k}{2}$$

where the constant k is the Fourier transform $M(j\omega)$ of the pseudo-impulse modulating signal. The output response of the band-pass filter is given by

$$h(t) = \int_{f_0-\Delta/2}^{f_0+\Delta/2} A\,\frac{k}{2}\exp[j\omega(t-T_d)]df + \int_{-f_0-\Delta/2}^{-f_0+\Delta/2} A\,\frac{k}{2}\exp[j\omega(t-T_d)]df$$

Integrating and substituting the limits, which is left to the student as an exercise, yields

$$h(t) = Ak\Delta\,\frac{\sin[2\pi\Delta(t-T_d)/2]}{2\pi\Delta(t-T_d)/2}\cos[\omega_c(t-T_d)] \qquad (I\text{–}26)$$

which is illustrated in Fig. I–25.

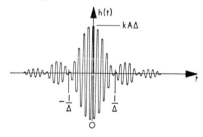

Fig. I–25 Output response of the ideal band-pass filter to a pseudo-impulse modulated carrier.

[I–6]

It is interesting to compare the result with the pseudo-impulse response of the ideal low-pass filter, given by Eq. (I–22), i.e.

$$h(t) = 2kAf_0 \frac{\sin[\omega_0(t-T_d)]}{\omega_0(t-T_d)}$$

If we substitute $f_0 = \Delta/2$ in this equation, then

$$h(t) = kA\Delta \frac{\sin[2\pi\Delta(t-T_d)/2]}{2\pi\Delta(t-T_d)/2} \qquad (I–27)$$

A comparison of Eqs. (I–26) and (I–27) shows that the response of the band-pass filter has an envelope which is the same shape as the response of the low-pass filter with cut-off frequency $f_0 = \Delta/2$, and for this reason the latter is referred to as the equivalent low-pass filter.

The general result can be evaluated by considering the band-pass filter to be defined as

$$H(j\omega) \text{ for } f_c-\Delta/2 < f < f_c+\Delta/2,$$

and $\qquad\qquad H(-j\omega) \text{ for } -f_c-\Delta/2 < f < -f_c+\Delta/2$

with $H(-j\omega) = H^*(j\omega)$, the necessary condition for the output response to be a real function. If the input is a modulated signal with a Fourier transform defined according to Eq. (I–25), then

$$g(t) = \int_{f_c-\Delta/2}^{f_c+\Delta/2} \tfrac{1}{2}M[j(\omega-\omega_c)]H(j\omega)\exp(j\omega t)df$$

$$+ \int_{-f_c-\Delta/2}^{-f_c+\Delta/2} \tfrac{1}{2}M[j(\omega+\omega_c)]H(-j\omega)\exp(j\omega t)df$$

Changing the limits of integration, the above equation becomes

$$g(t) = \int_{-\Delta/2}^{\Delta/2} \tfrac{1}{2}M(j\omega)H[j(\omega+\omega_c)\exp[j(\omega+\omega_c)t]df$$

$$+ \int_{-\Delta/2}^{\Delta/2} \tfrac{1}{2}M(j\omega)H[-j(\omega+\omega_c)]\exp[j(\omega-\omega_c)t]df$$

But since $H[j(\omega+\omega_c)]$ is the positive frequency transfer function shifted downwards to zero frequency, and $H[-j(\omega+\omega_c)]$ is the negative frequency counterpart shifted upwards to zero frequency (see Fig. I–26), then

$$H[j(\omega+\omega_c)] = H^*[-j(\omega+\omega_c)]$$

But the output $g(t)$ must be a real function; therefore we may write

$$g(t) = \text{Re}\left\{ \int_{-\Delta/2}^{\Delta/2} M(j\omega)H[j(\omega+\omega_c)]\exp(j\omega t)df \right\}\{\tfrac{1}{2}\exp(j\omega_c t)+\tfrac{1}{2}\exp(-j\omega_c t)\}$$

$$= \text{Re}\left\{ \int_{-\Delta/2}^{\Delta/2} M(j\omega)H[j(\omega+\omega_c)]\exp(j\omega t)df \right\}\cos\omega_c t \qquad (I\text{--}28)$$

The integral in Eq. (I–28) is the output response of the equivalent low-pass filter when the input is the modulating signal with Fourier transform $M(j\omega)$. Therefore, it is possible to determine the response of a band-pass filter to a modulated signal $m(t)\cos\omega_c t$ by evaluating the response of the equivalent low-pass filter to the modulating signal $m(t)$. It can be seen that when the transfer function of the band-pass filter is un-symmetrical about the centre frequency f_c, the equivalent low-pass filter is unrealistic because its output response would be complex. However, this is of no consequence, since we are able to ignore the imaginary part, due to the initial condition that the output $g(t)$ of the band-pass filter must be real.

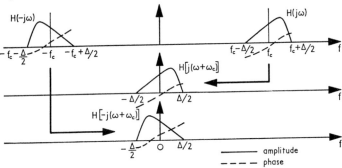

Fig. I–26 Band-pass and equivalent low-pass filter characteristics.

I–7 Power and Energy Spectra

So far, the Fourier analyses have been concerned with the amplitude spectra of time-varying functions; in the case of the periodic signal the line spectrum is defined in terms of the amplitude coefficient d_n, where-as the single event 'non-periodic' function has its spectrum defined as an amplitude density $F(j\omega)$. It is often more appropriate, and sometimes necessary, to work in terms of a power or energy spectrum.

I–7.1 *Power Spectrum of a Periodic Signal*

When a periodic function $f(t)$ defines either the voltage or current variations associated with a circuit, the power that would be dissipated in a

1 ohm resistor is known as the *normalised* power and the average value is given by

$$P_{av} = \frac{1}{T}\int_{-T/2}^{T/2} f^2(t)dt \quad \text{watts} \qquad (I–29)$$

Using Eq. (I–10), namely

$$f(t) = \sum_{n=-\infty}^{\infty} d_n \exp(jn\omega t)$$

and substituting for $f(t)$ in Eq. (I–29), gives

$$P_{av} = \frac{1}{T}\int_{-T/2}^{T/2} \left\{ \sum_{n=-\infty}^{\infty} d_n \exp(jn\omega t) \right\}^2 dt$$

An expansion of the integrand will consist of terms

$$[d_p \exp(jp\omega t)]\,[d_q \exp(jq\omega t)]$$

and the evaluation of the integral will depend upon the relation between p and q.

   (i) For $p \neq q$, the integral and hence the average power is zero (see the trigonometric relations on page 207).
   (ii) For $p = q$ and p positive (positive frequencies), the power contribution due to the pth component is

$$P_{av_p} = \frac{1}{T}\int_{-T/2}^{T/2} d_p^2\,(\cos p\omega t + j\sin p\omega t)^2 dt$$

$$= \frac{1}{T}\int_{-T/2}^{T/2} d_p^2(\cos p\omega t + 2j\cos p\omega t \sin p\omega t - \sin^2 p\omega t)dt$$

$$= d_p^2(\tfrac{1}{2}+0-\tfrac{1}{2}) = 0$$

Similarly, when $p = q$ and p is negative (negative frequencies),

$$P_{av_p} = 0$$

   (iii) For $p = -q$, i.e, a positive frequency component multiplied by its negative frequency counterpart, the power contribution is

$$P_{av\,|p|} = \frac{1}{T}\int_{-T/2}^{T/2} d_p\exp(jp\omega t)d_{-p}\exp(-jp\omega t)dt$$

$$= \frac{1}{T}\int_{-T/2}^{T/2} |d_p|^2 dt = |d_p|^2$$

Therefore the total power due to all the components is

$$P_{av} = d_0{}^2 + \sum_{n=-\infty}^{-1} |d_n|^2 + \sum_{n=1}^{\infty} |d_n|^2 \tag{I-30}$$

where $d_0{}^2$ is the power associated with the d.c. term.

Tutorial Exercise I-5. The two-sided representation of Eq. (I-30) is the result of Eq. (I-10) being used to define the periodic function $f(t)$. Repeat the analysis, but in place of Eq. (I-10), use the relation

$$f(t) = \sum_{n=1}^{\infty} 2\text{Re}[d_n \exp(jn\omega t)]$$

and show that the one-sided power spectrum is given by

$$P_{av} = d_0{}^2 + 2\sum_{n=1}^{\infty} |d_n|^2 \tag{I-31}$$

I-7.2 *Energy Spectrum of a Non-periodic Signal*

In Section I-4 the periodic function with $T \rightarrow \infty$ was used to define the non-periodic condition, for example a single pulse. We see from Eq. (I-29) that the average power will be zero and therefore it is meaningless to talk about the power spectrum of such a function; instead we can refer to the energy spectrum. The total energy E associated with the signal is

$$E = \int_{-\infty}^{\infty} f^2(t)dt$$

and using the Fourier transform results, Eqs. (I-17) and (I-21),

$$E = \int_{-\infty}^{\infty} f(t)dt \int_{-\infty}^{\infty} F(j\omega)\exp(j\omega t)df$$

$$= \int_{-\infty}^{\infty} F(j\omega)df \int_{-\infty}^{\infty} f(t)\exp(j\omega t)dt$$

$$= \int_{-\infty}^{\infty} F(j\omega)F(-j\omega)df$$

$$= \int_{-\infty}^{\infty} |F(j\omega)|^2 df \tag{I-32}$$

$$[\text{I-7}]$$

since $F(-j\omega) = F^*(j\omega)$, because $f(t)$ is a real function. This result is a particular case of *Parseval's Theorem*, which, in general, is concerned with the product of two functions, $f_1(t) f_2(t)$.

The factor $|F(j\omega)|^2$ has the dimensions (volts/Hz)^2, and when normalised for a 1 ohm load,

$$(\text{volts/Hz})^2 \equiv \text{watts/Hz}^2$$

$$\equiv \text{watt-seconds/Hz}$$

and therefore has the dimensions of an *energy spectral density*—a condition that is confirmed by inspecting Eq. (I–32).

Example I–10. Obtain an expression for the spectral energy density of a single rectangular pulse of amplitude V and duration t_p. Plot the energy density-frequency characteristic and compare it with an equivalent uniform spectrum having the same density at d.c. and accounting for the same energy in the pulse.

$$F(j\omega) = Vt_p \frac{\sin(\omega t_p/2)}{\omega t_p/2}$$

therefore, the energy spectral density is

$$|F(j\omega)|^2 = V^2 t_p{}^2 \left[\frac{\sin(\omega t_p/2)}{\omega t_p/2} \right]^2$$

which is illustrated in Fig. I–27.

The spectral density at d.c. is $V^2 t_p{}^2$ volt-seconds/Hz. If this value is held constant over a bandwidth $-f_e$ to $+f_e$, the total energy, i.e. the area under the curve, is $V^2 t_p{}^2 \times 2f_e$. Equating this to the pulse energy $V^2 t_p$, we have

$$V^2 t_p{}^2 \times 2f_e = V^2 t_p$$

therefore, $$f_e = \frac{1}{2t_p}$$

The equivalent square spectrum is shown in the figure.

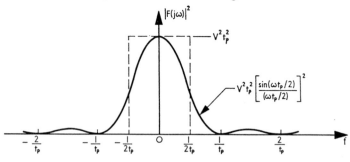

Fig. I–27 Energy density of a single rectangular pulse of amplitude V and duration t_p. (Broken curve represents a uniform spectrum accounting for the same energy.)

I-8 Convolution

In this appendix, we have been concerned mainly with determining the output response of a linear network by Fourier transform techniques and working in the frequency rather than the time domain. Quite often, even with the simplest input functions and networks, the inverse transform from $G(j\omega)$ to $g(t)$ involves an integral that is difficult and/or lengthy to evaluate. Take, for example, a rectangular pulse applied to a simple RC network; the output response is

$$g(t) = \int_{-\infty}^{\infty} Vt_p \frac{\sin(\omega t_p/2)}{\omega t_p/2} \frac{1}{1+j\omega CR} \exp(j\omega t) df$$

The evaluation of this integral should be avoided if there is an alternative and simpler method of deriving $g(t)$. For this particular example and for many others, the Convolution Integral provides such an alternative.

Consider the input function $f(t)$ to be identical to a function $f(\tau)$ which is made up of many impulses each defined as $f(n\delta\tau)$ (see Fig. I-28).

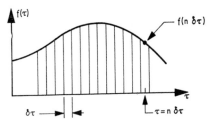

Fig. I-28 Impulse representation of a function $f(\tau)$.

The area of an impulse is $f(n\delta\tau)\,\delta\tau$—a unit impulse would have unit area. By definition, the response of a network to a unit impulse applied at $t = 0$ is $h(t)$, and applied at $t = \tau$, it is $h(t-\tau)$. The response to an impulse of area $f(n\delta\tau)\delta\tau$ applied at $t = n\delta\tau$ is therefore $f(n\delta\tau)$ $\delta\tau h(t-n\delta\tau)$. Hence, by superposition, the output response $g(t)$ is given by

$$g(t) \approx \sum_{n=-\infty}^{\infty} f(n\delta\tau)\delta\tau h(t-n\delta\tau)$$

and in the limit, $\delta\tau \to 0$, $n\delta\tau \to \tau$ and

$$g(t) = \int_{-\infty}^{\infty} f(\tau)h(t-\tau)d\tau \tag{I-33}$$

$$[\text{I-8}]$$

This result is a specific example of the Convolution Integral which, in general, is concerned with two functions $f_1(t)$ and $f_2(t)$. The following analysis deals with the general case and is presented here simply as an alternative derivation of the above expression.

Suppose that the Fourier transforms and inverse Fourier transforms, associated with the functions $f_1(t)$ and $f_2(t)$ are represented as

$$F_1(j\omega) = \mathfrak{F}[f_1(t)], \quad f_1(t) = \mathfrak{F}^{-1}[F_1(j\omega)] \qquad \text{(I–34a and b)}$$

$$F_2(j\omega) = \mathfrak{F}[f_2(t)], \quad f_2(t) = \mathfrak{F}^{-1}[F_2(j\omega)] \qquad \text{(I–35a and b)}$$

and that a third function $f(t)$ is defined by the relationships —

$$F_1(j\omega)F_2(j\omega) = \mathfrak{F}[f(t)], \quad f(t) = \mathfrak{F}^{-1}[F_1(j\omega)F_2(j\omega)] \quad \text{(I–36a and b)}$$

The question is: what is the relationship between $f(t)$, $f_1(t)$ and $f_2(t)$? Using Eq. (I–34a), but with the variable τ in place of t, we have

$$F_1(j\omega) = \int_{\tau=-\infty}^{\tau=\infty} f_1(\tau)\exp(-j\omega\tau)d\tau \qquad \text{(I–37)}$$

Also from Eq. (I–36b),

$$f(t) = \int_{f=-\infty}^{f=\infty} F_1(j\omega)F_2(j\omega)\exp(j\omega t)df \qquad \text{(I–38)}$$

and substituting Eq. (I–37) in Eq. (I–38),

$$f(t) = \int_{f=-\infty}^{f=\infty} \left[\int_{\tau=-\infty}^{\tau=\infty} f_1(\tau)\exp(-j\omega\tau)d\tau \right] F_2(j\omega)\exp(j\omega t)df$$

$$= \int_{\tau=-\infty}^{\tau=\infty} f_1(\tau) \left\{ \int_{f=-\infty}^{f=\infty} F_2(j\omega)\exp[j\omega(t-\tau)]df \right\}d\tau$$

Using the result of the Shift Theorem (see page 222),

$$f(t) = \int_{\tau=-\infty}^{\tau=\infty} f_1(\tau)f_2(t-\tau)d\tau \qquad \text{(I–39)}$$

which is identical to Eq. (I–33) when $f_2(t) = h(t)$.

The reason for Eq. (I–33) being known as the Convolution Integral is best understood by considering a graphical interpretation of the impulse response $h(t-\tau)$. When t is the variable and τ specifies a particular instant, $h(t-\tau)$ is identical to $h(t)$, but delayed in time by τ, as stated previously in the paragraph leading up to Eq. (I–33). But in Eqs. (I–33) and (I–39), τ is the variable and hence t specifies the particular instant; therefore $h(t-\tau)$, or $h(-\tau+t)$, plotted in the τ-plane is the same shape as the impulse response in the t-plane, except that it is folded over (or convoluted), see Fig. 1–29 which illustrates the impulse response of a *CR* network. Application of the Convolution Integral is straightforward provided that care is taken in choosing the limits of integration.

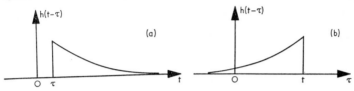

Fig. I–29 Impulse function $h(t-\tau)$ plotted with
(a) t as the variable, and
(b) τ as the variable.

For physically realizable filters, there cannot be an output before the input has been applied; if the time origin is arranged to coincide with the commencement of the input function the lower limit of the integral becomes 0 in place of $-\infty$; also the upper limit of $+\infty$ can be replaced by t when the output is required at some finite time. Hence, for the physically realizable network, we have

$$g(t) = \int_0^t f(\tau)h(t-\tau)d\tau \qquad (I\text{–}40)$$

Example I–11. To obtain the step response of a *RC* network.

$$f(t) = V, \quad t>0$$
$$f(t) = 0, \quad t<0$$
$$H(j\omega) = \frac{1}{1+j\omega CR}$$

Before the Convolution Integral can be applied, it is necessary to know the impulse response $h(t)$ of the network. Using the inverse Fourier transform

$$h(t) = \int_{-\infty}^{\infty} \frac{1}{1+j\omega CR} \exp(j\omega t)df$$

but the evaluation of this integral would defeat the present purpose of using the Convolution Integral, which is to avoid cumbersome integration. We note that for a function commencing at $t = 0$, the Fourier and Laplace transforms are identical, viz:

$$H(j\omega) = \int_0^{\infty} h(t)\exp(-j\omega t)dt$$

and

$$H(s) = \int_0^{\infty} h(t)\exp(-st)dt$$

That is to say,

$$H(j\omega) \equiv H(s) = \frac{1}{1+sCR}$$

$$[I\text{–}8]$$

Comprehensive tables of Laplace Transforms and operations are readily available,[9,10] and it is therefore possible to obtain the impulse response without undertaking the laborious task of evaluating the above integral. The moral is: consult tables of functions whenever possible and avoid unnecessary calculation. From the tables

$$h(t) = \frac{1}{RC} \exp(-t/RC)$$

or

$$h(t-\tau) = \frac{1}{RC} \exp[-(t-\tau)/RC]$$

We see from Fig. I–30 that $f(\tau) = 0$ for $\tau < 0$, and $h(t-\tau) = 0$ for $\tau > t$; therefore the limits of the Convolution Integral must be 0 and t and thus the output response at time t to a step input V is

$$g(t) = \int_0^t V \frac{1}{RC} \exp[-(t-\tau)/RC]d\tau$$

$$= V[1 - \exp(-t/RC)]$$

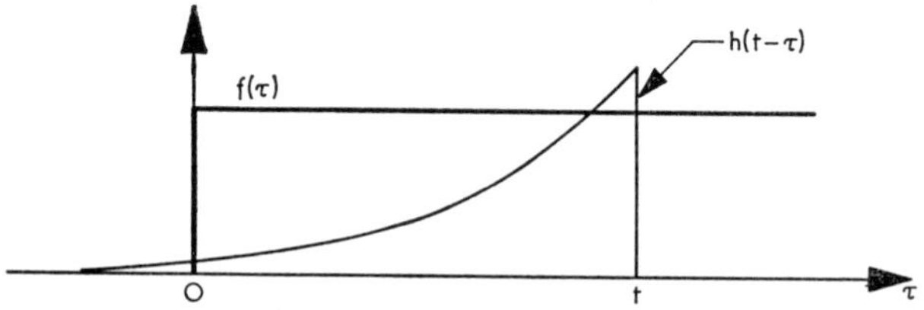

Fig. I–30 Calculation of the step response of a *RC* network by convolution.

Example I–12. To obtain the step response of an ideal low-pass filter.

$$f(t) = V, \ t > 0$$
$$f(t) = 0, \ t < 0$$
$$H(j\omega) = A \exp(-j\omega T_d), \ |f| < |f_0|$$
$$H(j\omega) = 0, \qquad\qquad |f| > |f_0|$$

The impulse response has been obtained previously in Example I–8; therefore

$$h(t-\tau) = 2Af_0 \frac{\sin[\omega_0(t-\tau-T_d)]}{\omega_0(t-\tau-T_d)}$$

Note that the existence of a precursor, i.e. $h(t) \neq 0$ for $t < 0$, rules out the use of Laplace Transform tables as an alternative method of obtaining $h(t)$.

From Fig. I–31 we see that $f(\tau) = 0$ for $\tau < 0$ and that the existence of the precursor of $h(t-\tau)$ means that all values of τ up to $+\infty$ must be included in the integration. Hence, the output response is given by

$$g(t) = \int_0^\infty 2AVf_0 \frac{\sin[\omega_0(t-\tau-T_d)]}{\omega_0(t-\tau-T_d)} d\tau$$

9. See References 2 and 3, page 228.
10. *Mathematical Handbook for Scientists and Engineers*, G. A. Korn and T. M. Korn (McGraw-Hill, 1961).

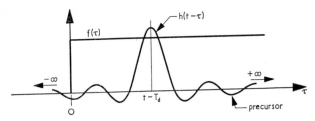

Fig. I–31 Calculation of the step response of an ideal low-pass filter by convolution.

Substituting $\omega_0(t-\tau-T_d) = x$,

$$g(t) = \frac{AV}{\pi} \int_{x=-\infty}^{x=\omega_0(t-T_d)} \frac{\sin x}{x} dx$$

But

$$\int_{-\infty}^{0} \frac{\sin x}{x} dx = \frac{\pi}{2}$$

Therefore $g(t) = \frac{AV}{2} \left\{ 1 + \frac{1}{\pi} \text{Si}[\omega_0(t-T_d)] \right\}$

(See also Example I–7, page 228.)

Example I–13. To determine the response of a $\sin x/x$ filter to a rectangular input pulse.

$$f(t) = V, \quad -t_p/2 < t < t_p/2$$

$$f(t) = 0, \quad \text{for all other } t$$

$$H(j\omega) = \frac{\sin(\omega t_p/2)}{\omega t_p/2} \exp(-j\omega T_d)$$

where the exponential term signifies a linear phase-frequency characteristic or constant time delay T_d. The impulse response $h(t)$ is given by the inverse Fourier transform

$$h(t) = \int_{-\infty}^{\infty} \frac{\sin(\omega t_p/2)}{\omega t_p/2} \exp[j\omega(t-T_d)] df \qquad \text{(I–41)}$$

There is no need to evaluate this integral or consult tables of Laplace transforms, since a similar form of integral is the well-known inverse Fourier transform of a rectangular pulse. We have, in the present application,

$$h(t) = \frac{1}{t_p}, \quad (T_d-t_p/2) < t < (T_d+t_p/2)$$

Note the t_p term in the denominator of this expression which arises on account of the difference between the expression for $H(j\omega)$ and that of Eq. (I–19), which defines $F(j\omega)$ for the rectangular pulse. Strictly speaking, $F_\delta(j\omega) = 1$ (the unit impulse Fourier transform) should be included in Eq. (I–41); we

[I–8]

therefore see that $h(t) = 1/t_p$ has, in fact, the dimensions of volts and not (seconds)$^{-1}$, as it may at first appear.

It can be seen from Fig. I–32 that the limits of the Convolution Integral will depend upon the relative positions of the impulse response $h(t-T_d-\tau)$ and the input $f(\tau)$; there are four possibilities:

(i) $(t-T_d) < -t_p/2,$ $g(t) = 0$

(ii) $-t_p/2 < (t-T_d) < t_p/2,$

$$g(t) = \int_{-t_p/2}^{t-T_d} V \frac{1}{t_p}\, d\tau = \frac{V}{2} + \frac{V(t-T_d)}{t_p}$$

(iii) $t_p/2 < (t-T_d) < 3t_p/2,$

$$g(t) = \int_{t-T_d-t_p}^{t_p/2} V \frac{1}{t_p}\, d\tau = \frac{V}{2} - \frac{V(t-T_d-t_p)}{t_p}$$

(iv) $(t-T_d) > 3t_p/2,$ $g(t) = 0$

The output response is triangular and has a peak amplitude when $(t-T_d) = t_p/2$

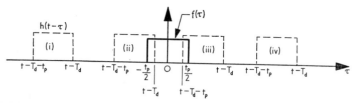

Fig. I–32 Calculation of the response of a $\sin x/x$ filter to a rectangular pulse, by convolution.

Further Tutorial Exercises

I–6   Use the Convolution Integral to obtain the output response of an ideal low-pass filter to a rectangular input. Check your result with Example I–7, page 228.

I–7   Use the Convolution Integral to show that the output response of an RC network to a rectangular pulse of amplitude V and duration t_p is

$$g(t) = V[1-\exp(-t/RC)], \quad 0 < t < t_p$$
$$= V\exp(-t/RC)[\exp(t_p/RC)-1], \qquad t > t_p$$

I–8   Show that the response of an ideal band-pass filter defined by the transfer function

$$H(j\omega) = A \exp(-j\omega T_d), \quad |f_c - \Delta/2| < |f| < |f_c + \Delta/2|$$
$$= 0 \text{ for all other frequencies,}$$

to a unit impulse function is

$$h(t) = 2\Delta \frac{\sin[\pi\Delta(t-T_d)]}{\pi\Delta(t-T_d)} \cos[\omega_c(t-T_d)]$$

Compare this expression with Eq. (I–26) and comment on the factor 2 difference.

I-9 Derive an expression for the Fourier transform of a 1μs rectangular pulse. Sketch the energy spectral density of the pulse and compare it with that corresponding to a triangular pulse of 2μs duration.

Comment on the bandwidths required for the transmission of these pulses and discuss briefly the effect of band-limiting such signals in order to reduce noise

$$Ans:\ F(j\omega)\bigg|_{\substack{\text{rectangular}\\\text{pulse}}} = V \times 10^{-6}\ \frac{\sin(\omega \times 10^{-6}/2)}{\omega \times 10^{-6}/2}$$

$$F(j\omega)\bigg|_{\substack{\text{triangular}\\\text{pulse}}} = V \times 10^{-6}\left[\frac{\sin(\omega \times 10^{-6}/2)}{\omega \times 10^{-6}/2}\right]^2$$

I-10 A rectangular pulse of 1 volt amplitude and 1μs duration is applied to a filter having a transfer function $H(j\omega)$ as shown in Fig. TE I-1. Evaluate the total energy of the output.

Hint: Compare the pulse duration with the cut-off characteristic of the filter.

$$Ans:\ 4 \times 10^{-9}\ V^2s.$$

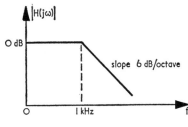

Fig. TE I-1.

I-11 A cosine squared pulse is defined by

$$f(t) = V \cos^2(\pi \times 10^6 t)\ \text{volts}$$

(See Fig. TE I-2.)

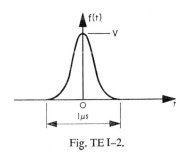

Fig. TE I-2.

[I-8]

Obtain an expression for its Fourier transform $F(j\omega)$ and hence determine the lowest frequency for which $F(j\omega) = 0$.

$$Ans:\ F(j\omega) = \frac{4V\pi^2\sin(\omega \times 10^{-6}/2)}{\omega(4\pi^2 - \omega^2 \times 10^{-12})}$$

$$f_{\min} = 2 \text{ MHz for } F(j\omega) = 0.$$

I–12   A raised cosine pulse is defined by

$$f(t) = \frac{V}{2}\left(1 + \cos\frac{2\pi t}{t_\mathrm{p}}\right),\ -t_\mathrm{p}/2 < t < t_\mathrm{p}/2$$

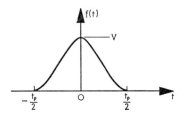

Fig. TE I–3.

Prove that the Fourier transform is

$$F(j\omega) = V\frac{\sin(\omega t_\mathrm{p}/2)}{\omega(1 - f^2 t_\mathrm{p}^2)}$$

When $t_\mathrm{p} = 5\mu s$ estimate the minimum bandwidth required for transmitting the pulse with negligible distortion.

Ans: 400 kHz.

Appendix II
PROBABILITY DISTRIBUTIONS

It is often convenient to analyse a communication system in terms of sine-wave inputs simply to allow deterministic representation. Whilst considerable information can be obtained from this form of analysis, there usually remains a number of unanswered questions; for example, a sine-wave analysis of an AM system would give no indication of the average depth of modulation when the input is natural speech—an extremely important consideration and one that can only be assessed from a knowledge of the statistical properties of the signal.

In Chapter 4 it is shown that noise is random and non-deterministic; it is also seen that a pseudo-deterministic form of representation enables the performance of analogue modulation methods in the presence of band-limited noise to be assessed. On the other hand, the analysis of digital systems, discussed in Chapter 8, cannot be carried out in terms of mean noise levels only, but requires additional information of the probability of the noise amplitude exceeding a level that, at the instant of sampling the signal, causes an incorrect decision to be made.

The purpose of this appendix is to introduce the concept of probability applied to electrical signals and noise and to show how the basic statistical definitions are related to the more familiar mean-square or r.m.s. amplitude. The treatment is non-rigorous and will only deal with the requirements of the main text. The reader is recommended to supplement this appendix with the more general and detailed accounts to be found in the literature.[1, 2, 3]

II–1 Statistics of a Single Variable
As an introduction, let us consider a variable that is observed on a finite number of occasions. For example, the output noise from a high-gain amplifier may be repeatedly sampled and a record made of its instantaneous amplitudes, either in the form of a dot-frequency dia-

1. *Principles and Applications of Random Noise Theory*, J. S. Bendat (Wiley, 1958).
2. *Introduction to Random Signals and Noise*, W. B. Davenport and W. L. Root (McGraw-Hill, 1958).
3. *An Introduction to Statistical Communication Theory*, D. Middleton (McGraw-Hill, 1960).

gram in which the number of dots is equal to the number of inspections, or as a stepped profile showing relative distributions; i.e. the ordinate of a step (or interval) $\varDelta v$ indicates the ratio of the number of occasions on which the instantaneous amplitude was found within the interval to the total number of samples taken; both of these forms of representation are shown in Fig. II–1. The range over which the amplitude can vary will be set by the saturation limits of the amplifier, say $\pm V$. In this example the variable is continuous over this range, i.e. it may assume

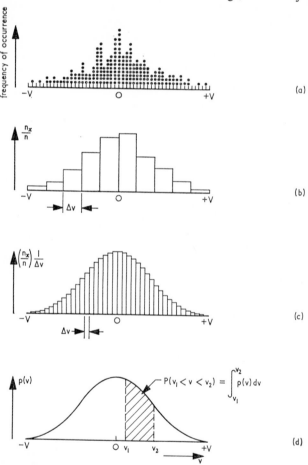

Fig. II–1 Amplitude distribution of the output noise of an amplifier.
(*a*) Dot frequency diagram comprising 200 observations.
(*b*) Relative distribution based on (*a*).
(*c*) Histogram made up of a considerably large number of observations.
(*d*) The continuous probability density function obtained when the number of observations tends to infinity.

any amplitude between the limits of $-V$ and $+V$. Variables can also be classified as *discrete*; for example, the well-known problem of picking an ace or aces from a pack of playing cards. Returning to the second diagram of Fig. II–1, the interval Δv must be chosen in accordance with the number of inspections made; if the latter is small and the interval is made too small, the step profile becomes discontinuous, i.e. no recordings in some of the intervals. But, since the variable is continuous over the range, such a probability profile would be misleading. Assuming that the difficulty can be avoided by correctly choosing the interval, there is also the disadvantage that the ordinates of the profile are dependent upon the interval—the smaller Δv the smaller the ordinate n_x/n. The histogram provides a method of overcoming this difficulty; the area of each rectangle represents the ratio n_x/n and therefore the ordinate, which is $(n_x/n)(1/\Delta v)$, remains constant as Δv is made smaller, provided that there is sufficient number of observations to avoid discontinuities. In the limit $\Delta v \to 0$ and $n \to \infty$, the stepped curve becomes continuous and the ordinate is referred to as the *probability density function $p(v)$*, which is

$$p(v) = \lim_{\substack{\Delta v \to 0 \\ n \to \infty}} \left(\frac{n_x}{n}\right) \left(\frac{1}{\Delta v}\right) \qquad \text{(II–1)}$$

and we see that $p(v) \geqslant 0$. The area under the curve between v_1 and v_2 gives the probability of the amplitude being found within that range, namely

$$P(v_1 < v < v_2) = \int_{v_1}^{v_2} p(v)dv \qquad \text{(II–2)}$$

In the example being considered the probability of the amplitude being found somewhere in the range $\pm V$ must be unity, hence

$$P(-V < v < +V) = \int_{-V}^{V} p(v)dv = 1 \qquad \text{(II–3)}$$

In general, the limits of the range are $+\infty$ and $-\infty$ and therefore

$$\int_{-\infty}^{\infty} p(v)dv = 1 \qquad \text{(II–4)}$$

It will be seen later in the appendix that if the r.m.s. amplitude of the noise is much smaller than V, the probability of the noise amplitude exceeding the saturation limit of the amplifier is extremely small and therefore V is to all intents and purposes equivalent to ∞.

[II–1]

Eq. (II–2) leads us to the cumulative distribution function $P(v)$, which is defined as

$$P(v_1) = \int_{-\infty}^{v_1} p(v)dv \tag{II–5}$$

$P(v_1)$ is the area under the curve from $-\infty$ to v_1 and is therefore the probability of the variable being less than v. It follows that

$$0 \leqslant P(v_1) \leqslant 1 \tag{II–6}$$

and
$$P(v_1 < v < v_2) = P(v_2) - P(v_1) \tag{II–7}$$

The histogram and probability density function specify the probability of the variable being within a particular portion of its range, and therefore they should provide sufficient information to enable the average value of the variable to be determined. Consider the data collected from a finite number of observations of a continuous variable. Let the range be divided into a large number of equal intervals Δv and the observations falling in the first interval to have an amplitude v_1, those in the second interval to have an amplitude v_2, and so on. If v_1 is observed N_1 times, v_2 observed. N_2 times, etc., the arithmetic mean is given by

$$\text{av } v = \frac{v_1 N_1 + v_2 N_2 + \ \ldots \ldots \ .v_n N_n}{N}$$

where $N = N_1 + N_2 + \ldots N_n$.

For a large number of observations the histogram becomes a continuous probability density function with

$$\frac{N_1}{N} = p(v_1)\Delta v, \qquad \frac{N_2}{N} = p(v_2)\Delta v, \text{ etc.}$$

and therefore

$$\text{av } v = \sum_{j=1}^{n} v_j \, p(v_j)\Delta v$$

In the limit $\Delta v \to 0$, and

$$\text{av } v = \int_{-\infty}^{\infty} v \, p(v) \, dv \tag{II–8}$$

The average value of v is also known as the *first moment* m_1, and similarly the average value of v^2 is the *second moment* m_2, i.e.

$$\text{av } v^2 = m_2 = \int_{-\infty}^{\infty} v^2 \, p(v) \, dv \tag{II–9}$$

In general, although we are concerned only with first and second moments, the nth order is given by

$$\text{av } v^n = m_n = \int_{-\infty}^{\infty} v^n \, p(v) \, dv \qquad (\text{II–10})$$

II–1.1 *Uniform (or Rectangular) Distribution*

Firstly, consider the non-electrical example of a pointer rotating at constant angular velocity. If the pointer is stopped at some arbitrary time, the probability of it being found within an elemental angle $d\theta$ is $p(\theta) \, d\theta$ and is independent of θ, i.e. $p(\theta)$ is a constant, k say. The range of angles is 2π and therefore from Eq. (II–4) we have

$$\int_0^{2\pi} p(\theta) \, d\theta = \int_0^{2\pi} k \, d\theta = 1$$

i.e. $$p(\theta) = \frac{1}{2\pi} \qquad (\text{II–11})$$

The probability density function is rectangular and the probability distribution given by

$$P(\theta) = \int_0^{\theta} p(\theta) \, d\theta = \frac{\theta}{2\pi} \qquad (\text{II–12})$$

(See Fig. II–2.)

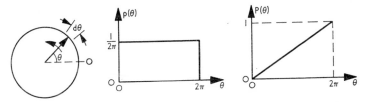

Fig. II–2 Probability density and cumulative distribution functions of a pointer rotating at constant angular velocity.

As a second example, let us refer to the quantization of an analogue signal in order that it may be converted into digital form (see Chapter 7). The quantizing step size Δv is small compared with the total range $+V$ to $-V$ and it may be assumed that, when the signal amplitude is within the range $v_j + (\Delta v)/2$ to $v_j - (\Delta v)/2$, the probability of it being found in an elemental interval dv is $p(v) \, dv$ and is independent of v, i.e. $p(v)$ is

[II–1]

constant over the step Δv. The error voltage, $\epsilon = v - v_\text{J}$, must also have uniform distribution, i.e. $p(\epsilon) = k$ say, and hence, using Eq. (II–4),

$$P[-(\Delta v)/2 < \epsilon < +(\Delta v)/2] = \int_{-\Delta v/2}^{\Delta v/2} k \, d\epsilon = 1$$

or
$$p(\epsilon) = \frac{1}{\Delta v} \qquad\qquad (II\text{–}13)$$

The mean square error which is used in assessing the quantization noise in PCM is

$$m_2 = \int_{-\Delta v/2}^{\Delta v/2} \epsilon^2 \, p(\epsilon) \, d\epsilon = \frac{(\Delta v)^2}{12} \qquad\qquad (II\text{–}14)$$

II–1.2 *Normal (or Gaussian) Distribution*

This is by far the most important probability distribution; it accounts for the statistical behaviour of random processes that are made up of an exceedingly large number of independent contributions; for example, the motion of electrons giving rise to thermal noise. The probability density function for a single variable such as the noise voltage appearing across a resistor is given by

$$p(v) = \frac{1}{\sigma\sqrt{2\pi}} \exp\left[-\frac{(v-\bar{v})^2}{2\sigma^2}\right] \qquad\qquad (II\text{–}15)$$

where $\bar{v} = m_1$, the average value, and σ is the *standard deviation,* a term which will now be explained. The shape of the probability density function which is illustrated in Fig. II–3 depends upon σ—the larger the standard deviation the flatter the curve and the greater the probability of amplitudes that are far removed from the mean value. The spread of the curve can be defined in many arbitrary ways, but the most commonly used definition is in terms of the mean square variation about a constant a, viz.

$$\text{spread, } s = \int_{-\infty}^{\infty} (v-a)^2 \, p(v) \, dv \qquad\qquad (II\text{–}16)$$

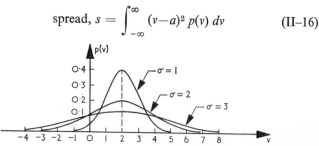

Fig. II–3 Normal (or Gaussian) probability density curves with standard deviation $\sigma = 1, 2$ and 3, and average value $m_1 = 2$.

Specifically, the spread measured about the mean value, i.e. $a = m_1$, is called the *variance* or *second central moment* μ_2,

$$\mu_2 = \int_{-\infty}^{\infty} (v-m_1)^2 \, p(v) \, dv$$

$$= \text{av } (v-m_1)^2 \qquad\qquad \text{(II–17)}$$

The square root of the variance is known as the standard deviation σ which is the factor appearing in the probability density function of Eq. (II–15). Expanding the squared term of Eq. (II–17) gives

$$\text{av } (v-m_1)^2 = \text{av } (v^2-2vm_1+m_1^2)$$

$$= \text{av } v^2-2m_1 \text{ av } v+m_1^2$$

$$= m_2-2m_1^2+m_1^2 = m_2-m_1^2$$

i.e. $$\mu_2 = \sigma^2 = m_2-m_1^2 \qquad\qquad \text{(II–18)}$$

When the variable is a fluctuating voltage or current, the second moment m_2 is a measure of the average power dissipated in a 1 ohm resistor. The d.c. power is equal to m_1^2 and hence from Eq. (II–18) the standard deviation σ is equal to the r.m.s. fluctuations about the average value. For the special case of $m_1 = 0$

$$p(v) = \frac{1}{\sigma\sqrt{2\pi}} \exp\left(-\frac{v^2}{2\sigma^2}\right)$$

$$m_1 = 0 \quad \text{and} \quad m_2 = \sigma^2$$

Although the probability density function of Eq. (II–15) has been stated without proof, the student may wish to substitute this expression into the integrals of Eqs. (II–8) and (II–9) and verify that av $v = \bar{v}$ and $\text{av}(v-\bar{v})^2 = \sigma^2$.

Returning to the original example of the output noise from a high-gain amplifier, suppose we wish to determine the probability of the amplitude exceeding a certain level or, alternatively, the proportion of the time for which the noise exceeds this level. This may be achieved by substituting Eq. (II–15) into Eq. (II–5), assuming that $\bar{v}$ and σ are known. The first may be measured on an instrument such as a moving-coil meter and the second, i.e. σ, by measuring the mean square amplitude m_2 and then substituting this value together with v into Eq. (II–18). The measuring instruments perform *time averages*; we may therefore write

$$[\text{II–1}]$$

$$m_1 = \overline{n(t)} \text{ or } <n(t)> = \lim_{T \to \infty} \frac{1}{T} \int_0^T n(t)\,dt \qquad \text{(II–19)}$$

and $\qquad m_2 = \overline{n^2(t)} \text{ or } <n^2(t)> = \lim_{T \to \infty} \frac{1}{T} \int_0^T n^2(t)\,dt \qquad \text{(II–20)}$

where $n(t)$ signifies the random variable and T is effectively the time constant of the instrument. Note the difference between the mathematical definition and the physical interpretation; in the limit, as $T \to \infty$, an acceptable approximation is realised by an instrument time constant that is very long compared with the fluctuation time of the noise source —this is considered in Appendix III under autocorrelation function. If m_1 and m_2 are independent of the actual time at which the measurements are made, the noise process is said to be *stationary*.

Another method of determining m_1, m_2 and σ, although not a practical proposition, is to sample simultaneously the outputs of a large number of identical noise sources (see Fig. II–4) and arithmetically average the

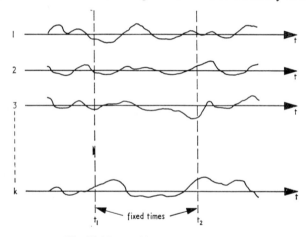

Fig. II–4 Ensemble of noise waveforms.

values of the samples to determine the statistical constants; this is referred to as the ensemble method and, since time averages are not involved, m_1 and m_2 are defined in the form of Eq. (II–8) and Eq. (II–9) respectively—assuming that the number of noise sources making up the ensemble approaches infinity. Again, the process is referred to as stationary if the statistics are time invariant, i.e. samples of the ensemble taken at a fixed time would result in the same probability density function as that from samples taken at any other time. Most randomly varying physical processes have identical time and ensemble averages and are referred to as *ergodic*; it follows that all ergodic processes are stationary; however, the converse is not true, i.e. a stationary process need not be ergodic.

We may now return to the problem posed earlier, that is, to determine the probability of the noise amplitude exceeding a certain level, v_1 say—assuming that the statistical constants m_1, m_2 and σ are known. From Eqs. (II–5) and (II–15), the probability distribution is

$$P(v_1) = \int_{-\infty}^{v_1} \frac{1}{\sigma\sqrt{2\pi}} \exp\left[-\frac{x^2}{2\sigma^2}\right] dx \qquad \text{(II–21)}$$

where $x = v - m_1$ (see Fig. II–5). Unfortunately, the integral cannot be evaluated in closed form, but must be expanded as a power series;

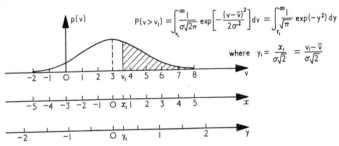

$$P(v > v_1) = \int_{v_1}^{\infty} \frac{1}{\sigma\sqrt{2\pi}} \exp\left[-\frac{(v-\bar{v})^2}{2\sigma^2}\right] dv = \int_{y_1}^{\infty} \frac{1}{\sqrt{\pi}} \exp(-y^2)\, dy$$

$$\text{where } y_1 = \frac{x_1}{\sigma\sqrt{2}} = \frac{v_1-\bar{v}}{\sigma\sqrt{2}}$$

Fig. II–5 Comparison of scales for a typical Gaussian probability density function, $m_1 = \bar{v} = 3$ and $\sigma = 2$.

however, this tedious process can be avoided, since the integral is to be found in tabulated form in various books of mathematical tables in which the variable is normalised in units of the standard deviation, i.e. by letting $x/\sigma\sqrt{2} = y$ say (see pages 186 and 286).

Hence

$$P(v_1) = \int_{-\infty}^{x_1/\sigma\sqrt{2}} \frac{1}{\sqrt{\pi}} \exp(-y^2)\, dy \qquad \text{(II–22)}$$

It can be seen from the symmetry of the probability density curve about $x = 0$ that

$$P(-|x_1|) = P(|x_1| < x < \infty) \qquad \text{(II–23)}$$

which may be written as

$$P(-\infty < x < -|x_1|) = P(|x_1| < x < \infty) \qquad \text{(II–24)}$$

Also

$$P(-|x_1| < x < |x_1|) = 1 - 2P(|x_1| < x < \infty) \qquad \text{(II–25)}$$
$$[\text{II–1}]$$

Using Eq. (II–2), we have

$$P(-|x_1|<x<|x_1|) = \frac{1}{\sqrt{\pi}} \int_{-y_1}^{y_1} \exp(-y^2)\, dy$$

$$= \frac{2}{\sqrt{\pi}} \int_{0}^{y_1} \exp(-y^2)\, dy \qquad \text{(II–26)}$$

where $y_1 = x_1/\sigma\sqrt{2}$ and $x_1 = v_1 - m_1$. The integral of Eq. (II–26) is called the *error function* and its abbreviation is

$$\text{erf } y_1 = \frac{2}{\sqrt{\pi}} \int_{0}^{y_1} \exp(-y^2)\, dy \qquad \text{(II–27)}$$

A short form table is to be found on page 286.

Example II–1. The output noise from a high-gain amplifier has an average value $m_1 = \overline{v(t)} = 1{\cdot}5$ V and a r.m.s. amplitude $\{\overline{v^2(t)}\}^{\frac{1}{2}} = 2{\cdot}25$ V. Determine the probability of the noise level being less than -3 V.

We wish to find the probability distribution $P(-3)$. Using Eq. (II–18) to find the standard deviation

$$\sigma^2 = m_2 - m_1{}^2 = (2{\cdot}25)^2 - (1{\cdot}5)^2 = 2{\cdot}81$$

therefore $\sigma = 1{\cdot}68$ V

From Eqs. (II–23), (II–25) and (II–26),

$$P(-3) = P(-\infty<v<-3) = \tfrac{1}{2}\{1-\text{erf } y_1\}$$

where, in this case, $y_1 = \dfrac{v_1-m_1}{\sigma\sqrt{2}} = \dfrac{-3-1{\cdot}5}{1{\cdot}68\sqrt{2}} = -1{\cdot}9$

The negative sign can be ignored, since the probability density function is symmetrical about $y = 0$ and from the tables on page 286, erf $(1{\cdot}9) = 0{\cdot}9928$, and therefore

$$P(-3) = 0{\cdot}0036$$

i.e. if the waveform were observed 10,000 times the amplitude would probably be less than -3 V for thirty-six of these observations.

II–2 Bivariate Distributions

There is often a requirement, as we shall see in the next appendix, to study the statistical behaviour of two variables that may or may not be independent of each other. As a simple illustration, consider two amplifiers in tandem; if the gain of the second amplifier is very low compared with the first, there will be a high degree of dependence (or *correlation*) between the two outputs (see Fig. II–6), but if the gain of the second amplifier is high the noise contributions at the outputs are from essentially different sources and the correlation will be small.

The degree of correlation cannot be assessed from a comparison of the single-variable constants m_1, m_2 and σ of each output. For example, the second amplifier could have unity gain and act solely as a d.c. level changer, thus altering the average value m_1; if the two waveforms were viewed on a double-beam oscilloscope, they would be seen to have identical fluctuation patterns—furthermore, the second moments m_2

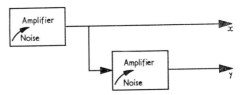

Fig. II–6 To illustrate the correlation of the noise outputs x and y from two amplifiers connected in cascade.

would be different, but the standard deviations would be equal. On the other hand, the noise outputs from two independent amplifiers could have identical constants m_1, m_2 and σ, but observation of their random fluctuations would show zero correlation between the two outputs. We must therefore look for another set of statistical constants that will serve as a measure of the interdependence of the variables.

Let us refer to the two variables as x and y. By a similar treatment to that for the single variable, a three- (in place of two-) dimensional histogram can be drawn up by dividing the respective ranges into intervals of Δx and Δy, repeatedly sampling the outputs and noting the frequency of occurrence of different pairs of x and y. If the number of observations is very large, the histogram profile becomes the probability density function $p(x, y)$. The probability that x will be found in the range $x \pm dx/2$ and y in the range $y \pm dy/2$ is

$$p(x, y) \, dx \, dy$$

By similar reasoning that led to Eq. (II–4), we have

$$\int_{-\infty}^{\infty} \int_{-\infty}^{\infty} p(x, y) \, dx \, dy = 1 \qquad (II–28)$$

Note that in this case the *volume* is normalised to unity, whereas for the single variable the *area* under the curve is equal to unity.

For the special case of two independent variables, it follows intuitively that

$$p(x, y) = p(x) \, p(y) \qquad (II–29)$$

and $\qquad p(x, y) \, dx \, dy = p(x) \, dx \, p(y) \, dy \qquad (II–30)$

$$[II–2]$$

The cumulative probability distribution is

$$P(x_1, y_1) = \int_{-\infty}^{x_1} \int_{-\infty}^{y_1} p(x, y) dx dy \qquad \text{(II–31)}$$

The general expression for the averages (or moments) of x and y is

$$m_{ij} = \text{av } x^i y^j = \int_{-\infty}^{\infty} \int_{-\infty}^{\infty} x^i y^j p(x, y) dx dy \qquad \text{(II–32)}$$

(See also the single variable definitions on page 254.)

For our purposes the following are important:

$$i = 1, j = 0, m_{10} = \text{av } x = \int_{-\infty}^{\infty} \int_{-\infty}^{\infty} x p(x, y) dx dy \qquad \text{(II–33)}$$

$$i = 0, j = 1, m_{01} = \text{av } y = \int_{-\infty}^{\infty} \int_{-\infty}^{\infty} y p(x, y) dx dy \qquad \text{(II–34)}$$

$$i = 1, j = 1, m_{11} = \text{av } xy = \int_{-\infty}^{\infty} \int_{-\infty}^{\infty} xy p(x, y) dx dy \qquad \text{(II–35)}$$

$$i = 2, j = 0, m_{20} = \text{av } x^2 = \int_{-\infty}^{\infty} \int_{-\infty}^{\infty} x^2 p(x, y) dx dy \qquad \text{(II–36)}$$

$$i = 0, j = 2, m_{02} = \text{av } y^2 = \int_{-\infty}^{\infty} \int_{-\infty}^{\infty} y^2 p(x, y) dx dy \qquad \text{(II–37)}$$

Similarly, the second central moments are defined as

$$\mu_{20} = \text{av } (x - m_{10})^2$$
$$= \text{av } (x^2 - 2x\, m_{10} + m_{10}^2)$$
$$= m_{20} - m_{10}^2 \qquad \text{(II–38)}$$

Similarly, $$\mu_{02} = m_{02} - m_{01}^2 \qquad \text{(II–39)}$$

and of particular interest, the *covariance* μ_{11} is

$$\mu_{11} = \text{av } (x - m_{10})(y - m_{01})$$
$$= \text{av } (xy - m_{10}\, y - m_{01}\, x + m_{10}\, m_{01})$$
$$= \text{av } xy - m_{10} \text{ av } y - m_{01} \text{ av } x + m_{10}\, m_{01}$$
$$= m_{11} - m_{10}\, m_{01} \qquad \text{(II–40)}$$

As a special case of independent variables

$$m_{11} = \text{av } xy = \text{av } x \text{ av } y = m_{10} \, m_{01}$$

and therefore $\qquad\qquad\qquad \mu_{11} = 0$

It will be seen in the following appendix that the joint average m_{11} and the covariance μ_{11} are used to define the *correlation* (or statistical dependence) of the two variables.

Tutorial Exercises

II-1 Substitute the probability density function of Eq. (II-15) into Eqs. (II-8) and (II-9), and hence verify that the variance μ_2 of the Gaussian distribution is equal to σ^2.

II-2 The envelope variations of band-limited Gaussian-distributed noise conform to the Rayleigh probability density function

$$p(r) = \frac{r}{N} \exp\left[-\frac{r^2}{2N} \right]$$

where N is the mean noise power (see Chapter 4). Use Eq. (II-9) and show that the second moment of r is $m_2 = 2N$. Comment on the factor 2.

II-3 A mean power of 1 mW is dissipated in a 50 ohm resistive load connected across the output terminals of a high-gain amplifier. The output is due solely to thermal agitation and shot noise within the amplifier (see Chapter 4). The d.c. level at the output is zero. What is the probability that the instantaneous voltage across the resistor lies between -1 and $+1$ V?

Ans: 0·99999214.

II-4 The joint probability density function of the two variables $x \geqslant 0$ and $y \geqslant 0$ is given by

$$p(x, y) = kxy\exp[-(x+y)]$$

Find the value of k such that $p(x,y)$ is properly normalised. Also calculate $P(x<1, y<2)$ and $P(x<1)$.

Ans: 1, 0·157, 0·264.

II-5 If v is normally distributed with mean m_1 and variance $\frac{1}{4}$, find
   (i) $P[x > (m_1 + \frac{1}{4})]$, and
   (ii) $P[(m_1 - \frac{1}{16}) < x < (m_1 + \frac{1}{16})]$

Ans: 2·77 × 10⁻⁷; 0·0995.

II-6 The probability density function defining the variable x is given by
   $$p(x) = k\exp(-2x), \qquad x \geqslant 0$$

Find the value of k such that $p(x)$ is properly normalised. Hence calculate the mean and variance of x.

Ans: 2; $\frac{1}{2}$, $\frac{1}{4}$.

[II-2]

Appendix III

AUTOCORRELATION AND POWER SPECTRA OF RANDOM FUNCTIONS

III–1 Correlation Functions*

The statistical dependence of two variables x and y can be expressed as the joint average of the variables, i.e.

$$m_{11} = \text{av } xy$$

(see Appendix II). Suppose that their averages are zero, av $x = $ av $y = 0$; then if x and y are independent

$$m_{11} = \text{av } xy = \text{av } x \text{ av } y = 0$$

At the other extreme, if x and y are identical,

$$m_{11} = \text{av } xy = \text{av } x^2 \text{ (or av } y^2) = m_2$$

We see, therefore, that for this particular example the joint average, which is known as the *cross-correlation function*, will range from zero to m_2 depending upon the degree of correlation between the two variables; this function may be defined in terms of ensemble averages,

$$m_{11} = \int_{-\infty}^{\infty} xy\, p(x,y)\, dx\, dy \tag{III–1}$$

or as a joint time average

$$R_{xy} = \lim_{T \to \infty} \frac{1}{T} \int_0^T x(t)y(t)\, dt \tag{III–2}$$

When the processes are ergodic $m_{11} = R_{xy}$ (see Appendix II).

Application of the cross-correlation function is to be found in the detection of very weak signals in the presence of noise. The receiving system consists of two identical but separate receivers; the outputs are multiplied together and integrated over a relatively long period, thus performing the process of cross-correlation. The signal components in the two branches will have a high degree of correlation, whereas the

---

* A comprehensive bibliography is to be found in *Correlation Techniques*, F. H. Lange (Iliffe, 1967).

noise will, to some extent, be uncorrelated. The output signal-to-noise ratio of the cross-correlator is therefore higher than the input signal-to-noise ratios. It is not proposed to describe the system in any further detail, as it has been introduced solely to illustrate the use of cross-correlation.

An equally important relationship is established by putting

$$y(t) = x(t+\tau)$$

to obtain the *autocorrelation function* $R(\tau)$, i.e. a process of cross-correlation between a single variable and itself displaced by a time τ. In terms of time averages,

$$R(\tau) = \lim_{T\to\infty} \frac{1}{T}\int_0^T x(t)x(t+\tau)dt \qquad \text{(III–3)}$$

and as an ensemble characteristic by the relation

$$m_{11} = \int_{-\infty}^{\infty}\int_{-\infty}^{\infty} x_1 x_2 p(x_1,x_2)dx_1 dx_2 \qquad \text{(III–4)}$$

where $p(x_1, x_2)$ is the joint probability density function obtained from pairs of values of x_1 and x_2 measured at time t_1 and $t_1+\tau$ respectively (see Fig. III–1). For an ergodic process $R(\tau) = m_{11}$.

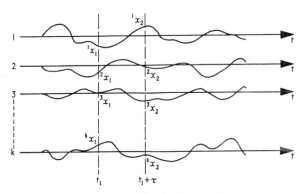

Fig. III–1 Ensemble of noise waveforms.

The following sections deal with the autocorrelation of periodic, aperiodic and continuous random functions. It is shown how the correlation function can describe the statistical properties of noise and also how it may be used to determine the spectral power densities of noise and random sequence digital signals.

[III–1]

III–1.1 *Autocorrelation of Periodic Functions*

We see from Eq. (III–3) that the autocorrelation function involves a time average and therefore, for the periodic function, $R(\tau)$ can be determined by averaging the product $x(t)\,x(t+\tau)$ over a complete period. Let us consider as an example a periodic sequence of rectangular pulses of unity mark-space ratio and period T (see Fig. III–2).

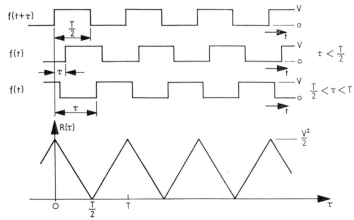

Fig. III–2 Autocorrelation of a periodic sequence of rectangular pulses of unity mark-space ratio.

When $\tau = 0$, the autocorrelation function is a maximum and given by

$$R(\tau) = R(0) = \frac{V^2}{2}$$

where V is the amplitude of the pulses. Also when

$$0 \leqslant \tau \leqslant T/2, \quad R(\tau) = \frac{1}{T}\int_{\tau}^{T/2} V^2 dt = V^2 \left(\frac{1}{2}-\frac{\tau}{T}\right)$$

and

$$T/2 \leqslant \tau \leqslant T, \quad R(\tau) = \frac{1}{T}\int_{T}^{\tau+T/2} V^2 dt = \frac{V^2}{T}\left(\tau-\frac{T}{2}\right)$$

The function is periodic and has a maximum value $R(0) = V^2/2$.

The autocorrelation function can also be expressed in terms of the line spectrum which has been defined previously in Appendix 1 (see Eq. (I–31), page 241). We may write

$$R(\tau) = \frac{1}{T}\int_{0}^{T} x(t)x(t+\tau)dt$$

$$= d_0{}^2 + 2\sum_{n=1}^{\infty} |d_n|^2 \cos\frac{2\pi n\tau}{T} \tag{III–5}$$

and when $\tau = 0$,

$$R(0) = \frac{1}{T}\int_0^T x^2(t)dt = d_0{}^2 + 2\sum_{n=1}^{\infty} |d_n|^2 \qquad \text{(III–6)}$$

which signifies that $R(0)$ is equal to the total power (see also Eq. (I–31). Furthermore, the autocorrelation function has the dimension of normalised watts (i.e. power dissipated in a 1 ohm load) when the dimension of the variable is either volts or amps. The significance of the correlation-power relationship is developed in the following sections.

III–1.2 *Autocorrelation of Aperiodic Functions*

The significant difference between a periodic and an aperiodic function is that the former contains an infinite amount of energy (since, for mathematical convenience, it exists over an infinite time scale) and finite average power, whereas the aperiodic function has finite energy and consequently zero average power, i.e.

$$P_{av} = \lim_{T \to \infty} \frac{1}{T}\int_{-T/2}^{T/2} x^2(t)dt = 0$$

and Energy,

$$E = \lim_{T \to \infty} \int_{-T/2}^{T/2} x^2(t)\,dt > 0$$

For this reason, the autocorrelation function is defined as

$$R(\tau) = \int_{-\infty}^{\infty} x(t)x(t+\tau)dt \qquad \text{(III–7)}$$

Since $x(t)$ contains finite energy—for example, it could define a single pulse—we may use the inverse Fourier transform

$$x(t+\tau) = \int_{-\infty}^{\infty} X(j\omega)\exp[\,j\omega(t+\tau)]df \qquad \text{(III–8)}$$

where $X(j\omega)$ is the Fourier transform of $x(t)$. Substituting into Eq. (III–7),

$$R(\tau) = \int_{t=-\infty}^{t=\infty} x(t)\left\{ \int_{f=-\infty}^{f=\infty} X(j\omega)\exp[\,j\omega(t+\tau)]\,df \right\}dt$$

Interchanging the order of integration,[1]

1. *Ancillary Mathematics*, H. S. W. Massey and H. Kestelman, Repeated (or Iterated) Integration, p. 437 (Pitman, 1959).

[III–1]

$$R(\tau) = \int_{f=-\infty}^{f=\infty} X(j\omega) \left\{ \int_{t=-\infty}^{t=\infty} x(t)\exp[j\omega(t+\tau)]dt \right\} df$$

$$= \int_{f=-\infty}^{f=\infty} X(j\omega)X(-j\omega)\exp(j\omega\tau)df$$

$$= \int_{-\infty}^{\infty} |X(j\omega)|^2 \exp(j\omega\tau)df \qquad \text{(III-9)}$$

since $X(j\omega) = X^*(-j\omega)$. In particular when $\tau = 0$

$$R(\tau) = R(0) = \int_{-\infty}^{\infty} |X(j\omega)|^2 \, df = \int_{-\infty}^{\infty} x^2(t) \, dt \qquad \text{(III-10)}$$

and the autocorrelation function is equal to the total energy of the function—note that the above equation is a restatement of Parseval's Theorem (see page 242).

III–1.3 *Autocorrelation of Continuous Random Functions*

We now come to the most important class of functions which includes noise processes and random sequence signals. Unlike an aperiodic function, $x(t)$ is continuous over the range $-\infty \leqslant t \leqslant \infty$ and the energy is infinite; hence, the integral

$$\int_{-\infty}^{\infty} x(t)\exp(-j\omega t)dt$$

does not converge and therefore the Fourier transform of $x(t)$ is meaningless. On the other hand, time averages do exist and we have

$$m_1 = \overline{x(t)} = \lim_{T\to\infty} \frac{1}{T}\int_{-T/2}^{T/2} x(t)dt \geqslant 0 \qquad \text{(III-11)}$$

$$m_2 = \overline{x^2(t)} = \lim_{T\to\infty} \frac{1}{T}\int_{-T/2}^{T/2} x^2(t)dt > 0 \qquad \text{(III-12)}$$

and therefore the autocorrelation function

$$R(\tau) = \overline{x(t)x(t+\tau)} = \lim_{T\to\infty} \frac{1}{T}\int_{-T/2}^{T/2} x(t)\, x(t+\tau)\, dt \qquad \text{(III-13)}$$

is meaningful. Similarly, the covariance (or *autocovariance* as it is known when dealing with the joint averages of a function multiplied by itself but displaced in time) is

$$\mu(\tau) = \text{av}[x(t)-m_1]\,[x(t+\tau)-m_1] \qquad \text{(III-14)}$$

(see Eq. (II–40), page 262). Combining Eqs. (III–13) and (III–14) and assuming the process to be stationary,

$$\mu(\tau) = R(\tau) - m_1^2 \qquad \text{(III–15)}$$

The following limiting conditions are of interest:

when $\tau = 0$, $\qquad R(\tau) = R(0) = \text{av } x^2 = m_2$

$$\mu(\tau) = \mu(0) = m_2 - m_1^2$$

and when $\tau \to \infty$ $\qquad R(\tau) \to m_1^2$ when $m_1 \neq 0$

$$R(\tau) \to 0 \text{ when } m_1 = 0$$

$\mu(\tau) \to 0$ regardless of whether m_1 is equal or not to zero.

Since the autocovariance, and the autocorrelation function when $m_1 = 0$, is finite and tends to zero as $\tau \to \infty$, we can write down the Fourier transform, with τ as the variable, as

$$\mathfrak{F}[\mu(\tau)] = \int_{-\infty}^{\infty} \mu(\tau) \exp(-j\omega\tau) d\tau$$

$$= S'(f) \quad \text{say.} \qquad \text{(III–16)}$$

$S'(f)$ must have the dimensions of watts/Hz, since $\mu(\tau)$ is in watts [Cf. $f(t)$—$F(j\omega)$ transforms in Appendix I]. The inverse Fourier transform is

$$\mu(\tau) = \int_{-\infty}^{\infty} S'(f) \exp(j\omega\tau) df \qquad \text{(III–17)}$$

and when $\tau = 0$,

$$\mu(\tau) = \mu(0) = \int_{-\infty}^{\infty} S'(f) df \qquad \text{(III–18)}$$

From the definition of the autocorrelation function, $R(0)$ is the total average power m_2 and therefore $\mu(0)$ is the total average power minus the d.c. power m_1^2. From Eq. (III–18) it follows that $S'(f)$ is a spectral power density, and since $\mu(0)$ is finite, $S'(f) \to 0$ as $f \to \infty$. We may also define the total power as

$$m_2 = R(0) = \int_{-\infty}^{\infty} S(f) df \qquad \text{(III–19)}$$

where $\qquad S(f) = S'(f) + m_1^2 \delta(f) \qquad \text{(III–20)}$

the impulse function $\delta(f)$ is zero except for $f = 0$ indicating that m_1^2

[III–1]

accounts for the d.c. power. Substituting Eqs. (III–15) and (III–20) into Eq. (III–17) gives

$$R(\tau)-m_1^2 = \int_{-\infty}^{\infty} [S(f)-m_1^2\delta(f)]\exp(j\omega\tau)df$$

$$= \int_{-\infty}^{\infty} S(f)\exp(j\omega\tau)df-m_1^2$$

and therefore,

$$R(\tau) = \int_{-\infty}^{\infty} S(f)\exp(j\omega\tau)df \qquad \text{(III–21)}$$

For a truly random process, the average value $m_1 = 0$, and therefore

$$\mu(\tau) = R(\tau) = \int_{-\infty}^{\infty} S(f)\exp(j\omega\tau)df \qquad \text{(III–22)}$$

and

$$S(f) = \int_{-\infty}^{\infty} \begin{bmatrix} R(\tau) \text{ or} \\ \mu(\tau) \end{bmatrix} \exp(-j\omega\tau)d\tau \qquad \text{(III–23)}$$

The Fourier transform pair of Eqs. (III–22) and (III–23), which state the relationship between the autocovariance, or autocorrelation, and spectral power density of the continuous process, is known as the *Wiener-Khintchine Theorem.*[*]

It has been stated previously that the Fourier integral involving $x(t)$ does not converge and therefore the Fourier transform $X(j\omega)$ is meaningless. This difficulty can be overcome by truncating the function such that

$$x(t) \equiv x_T(t), \ -T/2 \leqslant t \leqslant T/2$$

$$= 0, \ |t| > |T/2|$$

T can be very large and, provided that it is not taken to the limit, we have

$$X_T(j\omega) = \int_{-\infty}^{\infty} x_T(t)\exp(-j\omega t)dt \qquad \text{(III–24)}$$

where $X_T(j\omega)$ is meaningful. Consider the integral

$$\int_{-\infty}^{\infty} x_T(t)x_T(t+\tau)dt$$

* The original papers: Generalised Harmonic Analysis, N. Wiener, *Acta Math.*, Vol. 55, pp. 177–258, 1930, and Korrelationstheorie der Statistichen Stochastischen Prozesse, A. Khintchine, *Mathem. Annalen*, Vol. 109, pp. 604–15, 1933.

Carrying out the same operations that were used to establish Eq. (III–9), the integral becomes

$$\int_{-\infty}^{\infty} |X_T(j\omega)|^2 \exp(j\omega\tau)df$$

If $T \to \infty$, the expression becomes meaningless, but if the average is taken and the limiting condition applied, we have by previous definition, the autocorrelation function, namely

$$R(\tau) = \lim_{T \to \infty} \frac{1}{T} \int_{-\infty}^{\infty} |X_T(j\omega)|^2 \exp(j\omega\tau)df \qquad \text{(III–25)}$$

which is finite and meaningful. The order of the limiting process $\lim_{T \to \infty} \frac{1}{T}$ and the integration can only be interchanged if the integrand is a non-random function; a condition that is not satisfied in the integral of Eq. (III–25), since $X_T(j\omega)$ corresponds to the random function $x_T(t)$. However, the randomness can be removed by averaging over the ensemble of k members where $k \to \infty$, i.e.

$$\text{av}\{^kR(\tau)\} = \lim_{T \to \infty} \frac{1}{T} \int_{-\infty}^{\infty} \text{av}\{^k|X_T(j\omega)|^2\}\exp(j\omega\tau)df$$

The order of the limiting process and integration can now be interchanged; furthermore, for a stationary process the autocorrelation function is the same for every member of the ensemble and hence $\text{av}\{^kR(\tau)\} = R(\tau)$ therefore

$$R(\tau) = \int_{-\infty}^{\infty} \lim_{T \to \infty} \frac{\text{av}\{^k|X_T(j\omega)|^2\}}{T} \exp(j\omega\tau)df \qquad \text{(III–26)}$$

and comparing this equation with Eq. (III–21), we have

$$S(f) \equiv \lim_{T \to \infty} \frac{\text{av}\{^k|X_T(j\omega)|^2\}}{T} \qquad \text{(III–27)}$$

This result should in no way be surprising; it is known from previous analysis that $|X_T(j\omega)|^2$ has the dimensions of an energy spectral density and therefore its average value is a power density which has already been defined as $S(f)$. Its application is considered in section III–3.

III–2 Autocorrelation of Band-limited Noise

It is seen in Chapter 8 that the error probability associated with decision detection is related to the mean signal-to-noise ratio; however, no

[III–2]

indication of the frequency of occurrence of the errors is given, i.e. whether they are likely to occur singly or in groups. It is essential that this feature is known before one attempts to construct a code to protect the message.

The Wiener-Khintchine Theorem provides a useful method of analysis; if the spectral power density of the noise is known, the auto-correlation function can be determined by using Eq. (III–22), and this function will indicate the likelihood of successive noise samples being of similar amplitude.

III–2.1 *White Noise (occupying the Complete Spectrum)*

Let the one-sided spectral power density be μ watts/Hz. Using Eq. (III–22), the autocorrelation function

$$R(\tau) = \frac{\eta}{2} \int_{-\infty}^{\infty} \exp(j\omega\tau) df$$

where $\eta/2$ is the two-sided spectral power density. The integral represents the Fourier transform of a delta function $\delta(\tau)$, thus

$$R(\tau) = \frac{\eta}{2} \delta(\tau) \tag{III–28}$$

which means that $R(\tau)$ is zero except for $\tau \to 0$. This result is to be expected, since the noise has infinite bandwidth and its amplitude can assume a completely different value in an infinitesimally short period of time.

III–2.2 *Uniform Spectral Density Noise applied to an Ideal Low-pass Filter*

The spectral density of the output noise is

$$\frac{\eta}{2} | H(j\omega) |^2 \text{ watts} | \text{Hz}$$

and $$H(j\omega) = A, \quad -f_0 < f < f_0$$

Therefore, the autocorrelation function of the noise output is

$$R(\tau) = \frac{\eta}{2} A^2 \int_{-f_0}^{f_0} \exp(j\omega\tau) df$$

$$= \eta A^2 f_0 \frac{\sin 2\pi f_0 \tau}{2\pi f_0 \tau} \tag{III–29}$$

which is illustrated in Fig. III–3. We see that the correlation is very small when $\tau \geqslant 1/2f_0$, i.e. in a decision detection system, the errors would occur independently provided that the time between successive samples is not less than the reciprocal of twice the cut-off frequency of the network preceding the detector.

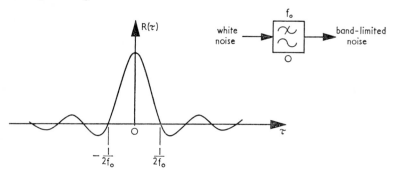

Fig. III–3 Autocorrelation function of band-limited noise.

It is interesting to note that the spectral power density of any process is always positive and has the dimension of watts/Hz, whereas the autocorrelation function has the dimension of watts, but can be either positive or negative depending upon the amplitude excursions of the time-varying process and the value of τ—it is always positive when $\tau = 0$, i.e. when it is equal to the total power.

III–3 Spectra of Random Sequence Digital Signals

Before embarking upon a mathematical analysis, it is considerably helpful to perform a qualitative investigation in order that a picture of the expected results may be visualised. First of all let us define the problem: we wish to establish the spectral power distribution of a random sequence digital signal that can be made up in any desired format, for example on-off, polar, bipolar, multilevel, etc.—we therefore seek a general analysis into which specific conditions can be substituted.

The reader is by now conversant with the line spectrum of a periodic sequence of pulses which, for ease of mathematical representation, commences at $t = -\infty$ and continues to $t = +\infty$. Suppose that the spectrum is viewed on the CRT display of a spectrum analyser. Now, in practice, the limits of infinity can never strictly apply; however, the line spectrum will be established in a relatively short period of time after connecting the input to the analyser. Next, consider the display that would correspond to a single pulse, i.e. an aperiodic signal. The average power is, of course, zero and the display would only be momentary;

[III–3]

but, if the CRT has an infinite or a long persistence characteristic, a continuous spectrum would be seen. Having considered the two extremes of a periodic sequence and a single pulse, we should next observe the spectrum of a random sequence continuous digital signal which, for simplicity of discussion, is assumed to be of on-off binary form. Without imposing too much strain on the elasticity of the imagination it is reasonable to assume that the display would comprise of both continuous and line spectra, but with amplitudes that vary randomly. For example, if there is an equal probability of each binary state being generated, it is conceivable that over some period of time the sequence could be made up of alternating states, thus giving rise to a well-defined line spectrum; on the other hand, there will be periods when the sequence consists of an isolated pulse.

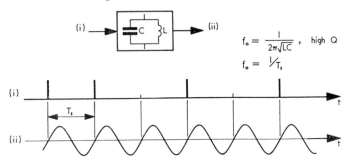

$$f_o = \frac{1}{2\pi\sqrt{LC}}, \quad \text{high } Q$$

$$f_o = 1/T_s$$

Fig. III–4 Excitation of an LC circuit with a random sequence of on-off impulses.

It is worth considering the existence of the line spectrum in more detail. If the random sequence is made up of impulses—the impulse is only used for ease of illustration and the conclusions to be drawn apply equally to finite duration pulses—and these are made to excite an *LC* circuit which is resonant at the fundamental period of the sequence (see Fig. III–4), then, provided that the *Q* of the circuit is high, the output will be a sine wave at the resonant frequency. Each impulse shock excites the circuit in such a way that the previously built-up oscillation is reinforced—in other words, there must be energy associated with that particular frequency, which implies that there is a discrete line in the spectrum. When the same resonant circuit is excited by a random sequence signal in polar form, i.e. 1 as a positive pulse and 0 as a negative pulse, succeeding pulses have a destructive influence upon an oscillation that may have been started previously; in fact, a steady-state sine wave at the resonant frequency of the *LC* circuit would not exist and it is therefore concluded that a polar sequence does not exhibit a line spectrum.

The following analysis[2] is specially chosen so that the continuous and line spectra of the random sequence signal may be separated.

A random time series of discrete numbers is

$$a_{-N} \ldots a_{-3}, a_{-2}, a_{-1}, a_0, a_1, a_2, a_3 \ldots a_n \ldots a_N$$

$$\rightarrow \text{time}$$

and an ensemble of binary sequences could be

av$\{^kM(t)\}$ = av$\{^kM(t)\}$				j		$j+q$	
1 . . . 0	1	1	0	1	0	1 . . . 1 . . . 0, $^1M(t)$
0 . . . 1	1	0	0	1	1	0 . . . 0 . . . 0, $^2M(t)$
1 . . . 1	0	0	1	1	0	0 . . . 0 . . . 1, $^3M(t)$
0 . . . 0	0	1	1	0	1	0 . . . 1 . . . 1, $^kM(t)$

where the superscript k attached to $M(t)$ indicates the member of the ensemble. Assuming the process to be ergodic, the ensemble average and time average of any one member are equal, i.e.

$$\text{av}\{^kM(t)\} = \overline{M(t)} \qquad \text{(III--30)}$$

Also the ensemble average does not depend upon the position in the sequence over which the average is taken (see above diagram).

For any member of the ensemble, the time average is

$$m_1 = \overline{M(t)} = \lim_{N \to \infty} \frac{1}{2N+1} \sum_{n=-N}^{N} a_n \qquad \text{(III--31)}$$

and the autocorrelation function is

$$R(q) = \text{av}\,(a_j a_{j+q}) = \lim_{N \to \infty} \frac{1}{2N+1} \sum_{j=-N}^{N} a_j a_{j+q} \qquad \text{(III--32)}$$

The autocorrelation function can also be defined in terms of an ensemble joint average, namely

$$R(q) = \text{av}\,(a_j a_{j+q}) \quad \text{for fixed } j; \qquad \text{(III--33)}$$

the average involves a summation of the k members.

2. Statistics of Regenerative Digital Transmission, W. R. Bennett, *Bell System Tech. J.*, Vol. 37, pp. 1501–42, November 1958.

[III–3]

The message in electrical form will consist of many pulses making up a continuous function of time, which for one member of the ensemble may be represented as

$$x(t) = \sum_{n=-\infty}^{\infty} a_n\, g(t-nT_s) \tag{III-34}$$

where $1/T_s = f_s$ the digit rate. For an on-off binary signal a_n is either unity or zero, in polar form $a_n = +1$ or -1 and so on; the function $g(t-nT_s)$ defines the pulse shape—the factor nT_s signifies the position of the nth pulse relative to the origin $t = 0$.

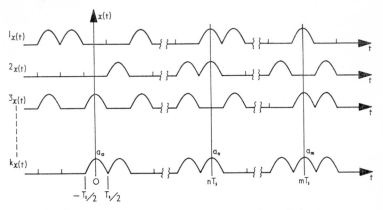

Fig. III–5 Ensemble of random sequences of on-off signals.

The reader who is meeting this form of representation for the first time must remember that the signal is random and the actual sequence cannot be defined in deterministic form. We see that there are two variables in Eq. (III–34), namely n and t. When $n = 0$, $x(t) = a_0 g(t)$ and t is a continuous variable over the interval $-T_s/2$ to $+T_s/2$. Another interpretation of the equation is that at fixed time t, $x(t)$ is made up of one coefficient weighted by the appropriate pulse factor $g(t-nT_s)$—both of these interpretations assume that the pulses do not overlap or that the overlapping is insignificant. This, however, is not an essential condition for the following analysis and, hence, in general, we have

$$x(t)\bigg|_{\substack{\text{fixed}\\\text{time}}} = a_{-n}g(t+nT_s)\ldots+a_{-1}g(t+T_s)+a_0g(t)+\ldots+a_ng(t-nT_s)$$

In the introductory remarks to this section it has been explained that a random sequence can exhibit a discrete line spectrum of varying amplitude. An average over the ensemble will, instead, show constant amplitude spectral characteristics, and it is for this reason that the ensemble rather than an individual member is to be studied. We can, of course, assume that whatever characteristics apply to the ensemble will also apply, *on average*, to any of its members.

At a fixed time t,

$$\text{av}\{{}^{k}x(t)\} = \text{av} \sum_{n=-\infty}^{\infty} a_{n}g(t-nT_{s})$$

Note that if the pulses do not overlap, $\text{av}\{{}^{k}x(t)\} = \text{av}\, a_{n}g(t-nT_{s})$.

The average of a sum is equal to the sum of the individual averages, therefore

$$\text{av}\{{}^{k}x(t)\}\bigg|_{\substack{\text{fixed}\\\text{time}}} = \sum_{n=-\infty}^{\infty} \text{av}\, a_{n}g(t-nT_{s})$$

$$= m_{1} \sum_{n=-\infty}^{\infty} g(t-nT_{s}) \tag{III-35}$$

The summation can be omitted from this expression if the pulses do not overlap.

If the above procedure is repeated at a fixed time $t+T_{s}$ in place of t,

$$\text{av}\{{}^{k}x(t+T_{s})\} = m_{1} \sum_{n=-\infty}^{\infty} g[t-(n-1)T_{s}]$$

and substituting $n' = n-1$, we have

$$\text{av}\{{}^{k}x(t+T_{s})\} \equiv \text{av}\{{}^{k}x(t)\} \tag{III-36}$$

Therefore the ensemble average not only varies over the pulse interval but it varies periodically at the pulse rate $1/T_{s}$. This is not at all surprising if reference is made to Fig. III-5; for the particular pulse shape illustrated, the ensemble average has a maximum value at the centre of the interval and is zero at the beginning and end of the time slot T_{s}. Since each member can occupy the complete time range from $-\infty$ to $+\infty$, let us represent the periodic ensemble average as a Fourier series expansion:

$$\text{av}\{{}^{k}x(t)\} = \sum_{r=-\infty}^{\infty} d_{r}\exp(jr\omega_{s}t)$$

$$[\text{III-3}]$$

where $\omega_s = 2\pi f_s = 2\pi/T_s$

and $$d_r = \frac{1}{T_s}\int^{T_s} \text{av}\{{}^k x(t)\}\exp(-jr\omega_s t)dt$$

$$= m_1 f_s \int_0^{T_s} \sum_{n=-\infty}^{\infty} g(t-nT_s)\exp(-jr\omega_s t)dt$$

Making the substitution $t-nT_s = u$

$$\exp(-jr\omega_s t) = \exp(-jr\omega_s u), \text{ since } \omega_s = 2\pi/T_s,$$

and $$\int_0^{T_s} \sum_{n=-\infty}^{\infty} = \int_{-nT_s}^{-(n-1)T_s} \sum_{n=-\infty}^{\infty} \equiv \int_{-\infty}^{\infty}$$

Therefore,

$$d_r = m_1 f_s \int_{-\infty}^{\infty} g(u)\exp(-jr\omega_s u)du$$

But the Fourier transform of the pulse $g(u)$ is

$$G(j\omega) = \int_{-\infty}^{\infty} g(u)\exp(-j\omega u)du$$

and we see that the integrals are identical when $\omega \equiv r\omega_s$. Hence, the coefficient d_r can be expressed as

$$d_r = m_1 f_s G(jr\omega_s)$$

For example, if $g(u)$ defines a unit-amplitude rectangular pulse of duration t_p, then

$$G(jr\omega_s) = t_p \frac{\sin(r\omega_s t_p/2)}{r\omega_s t_p/2}$$

Returning to the general analysis, the ensemble average becomes

$$\text{av}\{{}^k x(t)\} = m_1 f_s \sum_{r=-\infty}^{\infty} G(jr\omega_s)\exp(jr\omega_s t) \qquad \text{(III–37)}$$

This periodic characteristic of the ensemble average must give rise to a line spectrum at harmonic frequencies of f_s. To find such an expression was one of the original intentions, the other being to define the continuous part of the spectrum which we may now assume to be associated with an ensemble

$$\{^k y(t)\} = \{^k x(t)\} - \mathrm{av}\{^k x(t)\}$$

i.e. the original ensemble minus the average value to which we have attributed the line spectrum.

Let us evaluate the power density of the continuous spectrum by determining the average Fourier transform of the ensemble, $\mathrm{av}\{^k |Y_N(j\omega)|^2\}$ corresponding to a truncated set of members $\{^k y_N(t)\}$ and then substituting the result into Eq. (III–27), namely

$$S(f) = \lim_{N \to \infty} \frac{\mathrm{av}\{^k |Y_N(j\omega)|^2\}}{(2N+1)T_s}$$

where $(2N+1)\,T_s$ is the duration of the signal. A single member of the ensemble is

$$y_N(t) = \sum_{n=-N}^{N} (a_n - m_1) g(t - nT_s)$$

and its Fourier transform is

$$Y_N(j\omega) = \sum_{n=-N}^{N} (a_n - m_1) G(j\omega) \exp(-jn\omega T_s)$$

The square of the modulus of $Y_N(j\omega)$ must be written as

$$|Y_N(j\omega)|^2 = \sum_{n=-N}^{N} \sum_{m=-N}^{N} (a_n - m_1)(a_m - m_1)|G(j\omega)|^2 \exp[j\omega(n-m)T_s]$$

since conditions other than $n = m$ must be accounted for; it will be seen that this plays an important role in determining the spectra of signals for which there is interdigit correlation—for example, the polarity of a 1 digit in a bipolar sequence is determined by the polarity of the previous 1. The ensemble average of $|Y_N(j\omega)|^2$ is

$$\mathrm{av}\{^k |Y_N(j\omega)|^2\}$$

$$= \sum_{n=-N}^{N} \sum_{m=-N}^{N} \mathrm{av}[(a_n - m_1)(a_m - m_1)]|G(j\omega)|^2 \exp[j\omega(n-m)T_s]$$

$$[\text{III–3}]$$

The factors in the square brackets make up the autocovariance and, therefore, from Eq. (III–15) we have

$$\mathrm{av}[(a_n-m_1)(a_m-m_1)] = \mathrm{av}(a_n a_m)-m_1{}^2$$

The expression for the ensemble average of $|\,Y_N(j\omega)\,|^2$ needs rearranging and simplifying and in order that the operations may be facilitated, let $n-m = q$ say. Hence

$$\mathrm{av}\{^k|\,Y_N(j\omega)\,|^2\} = \sum_{n=-N}^{N}\sum_{q=n-N}^{N}[R(q)-m_1{}^2]\,|\,G(j\omega)\,|^2\exp(j\omega qT_s)$$

where $R(q) = \mathrm{av}\,(a_n a_m)$ and is the autocorrelation function of the ensemble. Further simplification can be achieved by rearranging the order of summation; it can be readily shown that

$$\sum_{n=-N}^{N}\sum_{q=n-N}^{n+N} \equiv \sum_{q=-2N}^{0}\sum_{n=-N}^{q+N} + \sum_{q=1}^{2N}\sum_{n=q-N}^{N}$$

Consequently,

$$\mathrm{av}\{^k|\,Y_N(j\omega)\,|^2\} = \sum_{q=-2N}^{0}(2N+1+q)[R(q)-m_1{}^2]\,|\,G(j\omega)\,|^2\exp(j\omega qT_s)$$

$$+ \sum_{q=1}^{2N}(2N+1-q)[R(q)-m_1{}^2]\,|\,G(j\omega)\,|^2\exp(j\omega qT_s)$$

and rearranging gives

$$\mathrm{av}\{^k|\,Y_N(j\omega)\,|^2\} = (2N+1)\,|\,G(j\omega)\,|^2\{R(0)-m_1{}^2$$

$$+ 2\sum_{q=1}^{2N}\left(1-\frac{q}{2N+1}\right)[R(q)-m_1{}^2]\cos\omega qT_s\}$$

Therefore,

$$\lim_{N\to\infty}\frac{\mathrm{av}\{^k|\,Y_N(j\omega)\,|^2\}}{(2N+1)T_s} \approx \frac{1}{T_s}\,|\,G(j\omega)\,|^2\{R(0)-m_1{}^2$$

$$+ 2\sum_{q=1}^{\infty}[R(q)-m_1{}^2]\cos\omega qT_s\} \tag{III–38}$$

which is the spectral power density of $\{^k y(t)\}$ and therefore accounts for the continuous spectrum of $\{^k x(t)\}$. The application of Eqs. (III–37) and (III–38), which represent the line and continuous spectra of a random sequence digital signal respectively, is considered in Chapter 8.

Tutorial Exercises

III–1 White noise is applied to an ideal low-pass filter of 1 kHz bandwidth. The output has a measured spectral power density of 0·5 mW/Hz. Show that the autocorrelation function of the output is

$$R(\tau) = 0\cdot 5 \frac{\sin(2\pi \times 10^3 \tau)}{2\pi \times 10^3 \tau} \text{ watts}$$

III–2 White noise of one-sided spectral power density kT watts/Hz is applied to a narrow-band amplifier centred on 100 kHz. The amplifier may be considered to have an ideal band-pass characteristic given by

$$|\,H(j\omega)\,| = 10^5, \quad 95 \text{ kHz} < f < 105 \text{ kHz}$$

If the noise figure of the amplifier is 3 dB, measured at 290°K, calculate the total noise power at the output.

Also obtain an expression for the autocorrelation function of the output noise.

The output noise voltage is to be sampled at intervals sufficiently apart to ensure almost independent samples. What should the time interval be? Boltzman's constant $k = 1\cdot 38 \times 10^{-23}$ J/°K

Ans: 8×10^{-7}W; $R(\tau) = 8 \times 10^{-7} \dfrac{\sin(\pi \times 10^4 \tau)}{\pi \times 10^4 \tau} \cos(2\pi \times 10^5 t)$

time interval > 100 µs.

III–3 The autocorrelation function of the output noise from a network is

$$R(\tau) = \frac{\eta \omega_0}{4} \exp[-\omega_0 |\tau|\,] \text{ watts}$$

where η is the one-sided spectral power density of the input noise and ω_0 is the frequency characteristic of the network. Show that the r.m.s. output is $\sqrt{\eta \omega_0/2}$ and also deduce the network.

Ans: RC network

III–4 A random sequence digital signal is made up of three pulse conditions, namely +1 V, zero and −1 V. There is equal probability of the positive and negative pulses occurring, $p(+1) = p(-1) = p/2$ say. Show that the two-sided spectral power density is given by

$$S(f) = pf_s |\,G(j\omega)\,|^2 \text{ watts} | \text{Hz}$$

where f_s is the digit rate and $G(j\omega)$ is the Fourier transform of the pulse.

Appendix IV
TABLES OF BESSEL FUNCTIONS

Table IV–1

n	$J_n(0.2)$	$J_n(0.4)$	$J_n(0.6)$	$J_n(0.8)$	$J_n(1.0)$	$J_n(1.25)$	$J_n(1.5)$	$J_n(1.75)$	$J_n(2.0)$	$J_n(2.5)$
0	+0·9900	+0·9604	+0·9120	+0·8463	+0·7652	+0·6459	+0·5118	+0·3690	+0·2239	−0·0484
1	+0·0995	+0·1960	+0·2867	+0·3688	+0·4401	+0·5106	+0·5579	+0·5802	+0·5767	+0·4971
2	+0·0050	+0·0197	+0·0437	+0·0758	+0·1149	+0·1711	+0·2321	+0·2940	+0·3528	+0·4461
3	+0·0002	+0·0013	+0·0044	+0·0102	+0·0196	+0·0369	+0·0610	+0·0918	+0·1289	+0·2166
4			+0·0003	+0·0010	+0·0025	+0·0059	+0·0118	+0·0209	+0·0340	+0·0738
5					+0·0002	+0·0007	+0·0018	+0·0038	+0·0070	+0·0195
6							+0·0002	+0·0006	+0·0012	+0·0042
7								+0·0001	+0·0002	+0·0008
8										+0·0001

Signal Processing, Modulation and Noise

Table

n	$J_n(1)$	$J_n(2)$	$J_n(3)$	$J_n(4)$	$J_n(5)$	$J_n(6)$	$J_n(7)$	$J_n(8)$	$J_n(9)$	J_n
0	+0·7652	+0·2239	−0·2601	−0·3971	−0·1776	+0·1506	+0·3001	+0·1717	−0·0903	−0
1	+0·4400	+0·5767	+0·3391	−0·0660	−0·3276	−0·2767	−0·0047	+0·2346	+0·2453	+0
2	+0·1149	+0·3528	+0·4861	+0·3641	+0·0466	−0·2429	−0·3014	−0·1130	+0·1448	+0
3	+0·0196	+0·1289	+0·3091	+0·4302	+0·3648	+0·1148	−0·1676	−0·2911	−0·1809	+0
4	+0·0025	+0·0340	+0·1320	+0·2811	+0·3912	+0·3576	+0·1578	−0·1054	−0·2655	−0
5	+0·0002	+0·0070	+0·0430	+0·1321	+0·2611	+0·3621	+0·3479	+0·1858	−0·0550	−0
6		+0·0012	+0·0114	+0·0491	+0·1310	+0·2458	+0·3392	+0·3376	+0·2043	−0
7		+0·0002	+0·0025	+0·0152	+0·0534	+0·1296	+0·2336	+0·3206	+0·3275	+0
8			+0·0005	+0·0040	+0·0184	+0·0565	+0·1280	+0·2235	+0·3051	+0
9				+0·0009	+0·0055	+0·0212	+0·0589	+0·1263	+0·2149	+0
10				+0·0002	+0·0015	+0·0070	+0·0235	+0·0608	+0·1247	+0
11					+0·0004	+0·0020	+0·0083	+0·0256	+0·0622	+0
12						+0·0005	+0·0027	+0·0096	+0·0274	+0
13						+0·0001	+0·0008	+0·0033	+0·0108	+0
14							+0·0002	+0·0010	+0·0039	+0
15								+0·0003	+0·0013	+0
16									+0·0004	+0
17									+0·0001	+0
18										+0
19										
20										
21										
22										
23										
24										
25										

For more detailed information see, *Bessel Functions*, Enzo Cambi (Dover, 1948).

$J_n(11)$	$J_n(12)$	$J_n(13)$	$J_n(14)$	$J_n(15)$	$J_n(16)$	$J_n(17)$	$J_n(18)$	$J_n(19)$	$J_n(20)$
1712	+0·0477	+0·2069	+0·1711	−0·0142	−0·1749	−0·1699	−0·0134	+0·1466	+0·1670
1768	−0·2234	−0·0703	+0·1334	+0 2051	+0·0904	−0·0977	−0·1880	−0·1057	+0·0668
1390	−0·0849	−0·2177	−0·1520	+0·0416	+0·1862	+0·1584	−0·0075	−0·1578	−0·1603
2273	+0·1951	+0·0033	−0·1768	−0·1940	−0·0438	+0·1349	+0·1863	+0·0725	−0·0989
0150	+0·1825	+0·2193	+0·0762	−0·1192	−0·2026	−0·1107	+0·0696	+0·1806	+0·1307
2383	−0·0735	+0·1316	+0·2204	+0·1305	−0·0575	−0·1870	−0·1554	+0·0036	+0·1512
2016	−0·2437	−0·1180	+0·0812	+0·2061	+0·1667	+0·0007	−0·1560	−0·1788	−0·0551
0184	−0·1703	−0·2406	−0·1508	+0·0345	+0·1825	+0·1875	+0·0514	−0·1165	−0·1842
2250	+0·0451	−0·1410	−0·2320	−0·1740	−0·0070	+0·1537	+0·1959	+0·0929	−0·0739
3089	+0·2304	+0·0670	−0·1143	−0·2200	−0·1895	−0·0429	+0·1228	+0·1947	+0·1251
2804	+0·3005	+0·2338	+0·0850	−0·0901	−0·2062	−0·1991	−0·0732	+0·0916	+0·1865
2010	+0·2704	+0·2927	+0·2357	+0·1000	−0·0682	−0·1914	−0·2041	−0·0984	+0·0614
1216	+0·1953	+0·2615	+0·2855	+0·2367	+0·1124	−0·0486	−0·1762	−0·2055	−0·1190
0643	+0·1201	+0·1901	+0·2536	+0·2787	+0·2368	+0·1228	−0·0309	−0·1612	−0·2041
0304	+0·0650	+0·1188	+0·1855	+0·2464	+0·2724	+0·2364	+0·1316	−0·0151	−0·1464
0130	+0·0316	+0·0656	+0·1174	+0·1813	+0·2399	+0·2666	+0·2356	+0·1389	−0·0008
0051	+0·0140	+0·0327	+0·0661	+0·1162	+0·1775	+0·2340	+0·2611	+0·2345	+0·1452
0019	+0·0057	+0·0149	+0·0337	+0·0665	+0·1150	+0·1739	+0·2286	+0·2559	+0·2331
0006	+0·0022	+0·0063	+0·0158	+0·0346	+0·0668	+0·1138	+0·1706	+0·2235	+0·2511
0002	+0·0008	+0·0025	+0·0068	+0·0166	+0·0354	+0·0671	+0·1127	+0·1676	+0·2189
	+0·0003	+0·0009	+0·0028	+0·0074	+0·0173	+0·0362	+0·0673	+0·1116	+0·1647
		+0·0003	+0·0010	+0·0031	+0·0079	+0·0180	+0·0369	+0·0675	+0·1106
		+0·0001	+0·0004	+0·0012	+0·0034	+0·0084	+0·0187	+0·0375	+0·0676
			+0·0001	+0·0004	+0·0013	+0·0037	+0·0089	+0·0193	+0·0381
				+0·0005	+0·0015	+0·0039	+0·0093	+0·0199
				+0·0002	+0·0006	+0·0017	+0·0042	+0·0098

Appendix V
NORMAL ERROR FUNCTION TABLES

$$\operatorname{erf} x = \frac{2}{\sqrt{\pi}} \int_0^x \exp(-y^2)\, dy$$

x	erf x	x	erf x	x	erf x	x	erf x
0·00	0.000 000	0·40	0·428 392	0·80	0·742 101	1·20	0·910 314
0·01	0·011 283	0·41	0·437 969	0·81	0·748 003	1·21	0·912 956
0·02	0·022 565	0·42	0·447 468	0·82	0·753 811	1·22	0·915 534
0·03	0·033 841	0·43	0·456 887	0.83	0·759 524	1·23	0·918 050
0·04	0·045 111	0·44	0·466 225	0·84	0·765 143	1·24	0·920 505
0·05	0·056 372	0·45	0·475 482	0·85	0·770 668	1·25	0·922 900
0·06	0·067 622	0·46	0·484 655	0·86	0·776 100	1·26	0·925 236
0·07	0·078 858	0·47	0·493 745	0·87	0·781 440	1·27	0·927 514
0·08	0·090 078	0·48	0·502 750	0·88	0·786 687	1·28	0·929 734
0·09	0·101 281	0·49	0·511 668	0·89	0·791 843	1·29	0·931 899
0·10	0·112 463	0·50	0·520 500	0·90	0·796 908	1·30	0·934 008
0·11	0·123 623	0·51	0·529 244	0·91	0·801 883	1·31	0·936 063
0·12	0·134 758	0·52	0·537 899	0·92	0·806 768	1·32	0·938 065
0·13	0·145 867	0·53	0·546 464	0·93	0·811 564	1·33	0·940 015
0·14	0·156 947	0·54	0·554 939	0·94	0·816 271	1·34	0·941 914
0·15	0·167 996	0·55	0·563 323	0·95	0·820 891	1·35	0·943 762
0·16	0·179 012	0·56	0·571 616	0·96	0·825 424	1·36	0·945 561
0·17	0·189 992	0·57	0·579 816	0·97	0·829 870	1·37	0.947 312
0·18	0·200 936	0·58	0·587 923	0·98	0·834 232	1·38	0·949 016
0·19	0·211 840	0·59	0·595 936	0·99	0·838 508	1·39	0·950 673
0·20	0·222 703	0·60	0·603 856	1·00	0·842 701	1·40	0·952 285
0·21	0·233 522	0·61	0·611 681	1·01	0·846 810	1·41	0·953 852
0·22	0·244 296	0·62	0.619 411	1·02	0.850 838	1·42	0·955 376
0·23	0·255 023	0·63	0·627 046	1·03	0·854 784	1·43	0·956 857
0·24	0·265 700	0·64	0·634 586	1·04	0·858 650	1·44	0·958 297
0·25	0·276 326	0·65	0.642 029	1·05	0·862 436	1·45	0·959 695
0·26	0·286 900	0·66	0·649 377	1·06	0·866 144	1·46	0·961 054
0·27	0·297 418	0·67	0·656 628	1·07	0·869 773	1·47	0·962 373
0·28	0·307 880	0·68	0·663 782	1·08	0·873 326	1·48	0·963 654
0·29	0·318 283	0·69	0·670 840	1·09	0·876 803	1·49	0·964 898
0·30	0·328 627	0·70	0·677 801	1·10	0·880 205	1·50	0·966 105
0·31	0·338 908	0·71	0·684 666	1·11	0·883 533	1·51	0·967 277
0·32	0·349 126	0·72	0.691 433	1·12	0·886 788	1·52	0·968 413
0·33	0·359 279	0·73	0·698 104	1·13	0·889 971	1·53	0·969 516
0·34	0·369 365	0·74	0·704 678	1·14	0·893 082	1·54	0·970 586
0·35	0·379 382	0·75	0·711 156	1·15	0·896 124	1·55	0·971 623
0·36	0·389 330	0·76	0·717 537	1·16	0·899 096	1·56	0·972 628
0·37	0·399 206	0·77	0·723 822	1·17	0·902 000	1·57	0·973 603
0·38	0·409 009	0·78	0·730 010	1·18	0·904 837	1·58	0·974 547
0·39	0·418 739	0·79	0·736 103	1·19	0·907 608	1·59	0·975 462

x	erf x	x	erf x	x	erf x	x	erf x
1·60	0·976 348	2·10	0·997 021	2·60	0·999 764	3·10	0·999 988 35
1·61	0·977 207	2·11	0·997 155	2·61	0·999 777	3·11	0·999 989 08
1·62	0·978 038	2·12	0·997 284	2·62	0·999 789	3·12	0·999 989 77
1·63	0·978 843	2·13	0·997 407	2·63	0·999 800	3·13	0·999 990 42
1·64	0·979 622	2·14	0·997 525	2·64	0·999 811	3·14	0·999 991 03
1·65	0·980 376	2·15	0·997 639	2·65	0·999 822	3·15	0·999 991 60
1·66	0·981 105	2·16	0·997 747	2·66	0·999 831	3·16	0·999 992 14
1·67	0·981 810	2·17	0·997 851	2·67	0·999 841	3·17	0·999 992 64
1·68	0·982 493	2·18	0·997 951	2·68	0·999 849	3·18	0·999 993 11
1·69	0·983 153	2·19	0·998 046	2·69	0·999 858	3·19	0·999 993 56
1·70	0·983 790	2·20	0·998 137	2·70	0·999 866	3·20	0·999 993 97
1·71	0·984 407	2·21	0·998 224	2·71	0·999 873	3·21	0·999 994 36
1·72	0·985 003	2·22	0·998 308	2·72	0·999 880	3·22	0·999 994 73
1·73	0·985 578	2·23	0·998 388	2·73	0·999 887	3·23	0·999 995 07
1·74	0·986 135	2·24	0·998 464	2·74	0·999 893	3·24	0·999 995 40
1·75	0·986 672	2·25	0·998 537	2·75	0·999 899	3·25	0·999 995 70
1·76	0·987 190	2·26	0·998 607	2·76	0·999 905	3·26	0·999 995 98
1·77	0·987 691	2·27	0·998 674	2·77	0·999 910	3·27	0·999 996 24
1·78	0·988 174	2·28	0·998 738	2·78	0·999 916	3·28	0·999 996 49
1·79	0·988 641	2·29	0·998 799	2·79	0·999 920	3·29	0·999 996 72
1·80	0·989 091	2·30	0·998 857	2·80	0·999 925	3·30	0·999 996 94
1·81	0·989 525	2·31	0·998 912	2·81	0·999 929	3·31	0·999 997 15
1·82	0·989 943	2·32	0·998 966	2·82	0·999 933	3·32	0·999 997 34
1·83	0·990 347	2·33	0·999 016	2·83	0·999 937	3·33	0·999 997 51
1·84	0·990 736	2·34	0·999 065	2·84	0·999 941	3·34	0·999 997 68
1·85	0·991 111	2·35	0·999 111	2·85	0·999 944	3·35	0·999 997 838
1·86	0·991 472	2·36	0·999 155	2·86	0·999 948	3·36	0·999 997 983
1·87	0·991 821	2·37	0·999 197	2·87	0·999 951	3·37	0·999 998 120
1·88	0·992 156	2·38	0·999 237	2·88	0·999 954	3·38	0·999 998 247
1·89	0·992 479	2·39	0·999 275	2·89	0·999 956	3·39	0·999 998 367
1·90	0·992 790	2·40	0·999 311	2·90	0·999 959	3·40	0·999 998 478
1·91	0·993 090	2·41	0·999 346	2·91	0·999 961	3·41	0·999 998 582
1·92	0·993 378	2·42	0·999 379	2·92	0·999 964	3·42	0·999 998 679
1·93	0·993 656	2·43	0·999 411	2·93	0·999 966	3·43	0·999 998 770
1·94	0·993 923	2·44	0·999 441	2·94	0·999 968	3·44	0·999 998 855
1·95	0·994 179	2·45	0·999 469	2·95	0·999 970	3·45	0·999 998 934
1·96	0·994 426	2·46	0·999 497	2·96	0·999 972	3·46	0·999 999 008
1·97	0·994 664	2·47	0·999 523	2·97	0·999 973	3·47	0·999 999 077
1·98	0·994 892	2·48	0·999 547	2·98	0·999 975	3·48	0·999 999 141
1·99	0·995 111	2·49	0·999 571	2·99	0·999 977	3·49	0·999 999 201
2·00	0·995 322	2·50	0·999 593	3·00	0·999 977 91	3·50	0·999 999 257
2·01	0·995 525	2·51	0·999 614	3·01	0·999 979 26	3·51	0·999 999 309
2·02	0·995 719	2·52	0·999 635	3·02	0·999 980 53	3·52	0·999 999 358
2·03	0·995 906	2·53	0·999 654	3·03	0·999 981 73	3·53	0·999 999 403
2·04	0·996 086	2·54	0·999 672	3·04	0·999 982 86	3·54	0·999 999 445
2·05	0·996 258	2·55	0·999 689	3·05	0·999 983 92	3·55	0·999 999 485
2·06	0·996 423	2·56	0·999 706	3·06	0·999 984 92	3·56	0·999 999 521
2·07	0·996 582	2·57	0·999 722	3·07	0·999 985 86	3·57	0·999 999 555
2·08	0·996 734	2·58	0·999 736	3·08	0·999 986 74	3·58	0·999 999 587
2·09	0·996 880	2·59	0·999 751	3·09	0·999 987 57	3·59	0·999 999 617

x	erf x	x	erf x	x	erf x	x	erf x
3·60	0·999 999 644	3·70	0·999 999 833	3·80	0·999 999 923	3·90	0·999 999 965
3·61	0·999 999 670	3·71	0·999 999 845	3·81	0·999 999 929	3·91	0·999 999 968
3·62	0·999 999 694	3·72	0·999 999 857	3·83	0·999 999 934	3·92	0·999 999 970
3·63	0·999 999 716	3·73	0·999 999 867	3·83	0·999 999 939	3·93	0·999 999 973
3·64	0·999 999 736	3·74	0·999 999 877	3·84	0·999 999 944	3·94	0·999 999 975
3·65	0·999 999 756	3·75	0·999 999 886	3·85	0·999 999 948	3·95	0·999 999 977
3·66	0·999 999 773	3·76	0·999 999 895	3·86	0·999 999 952	3·96	0·999 999 979
3·67	0·999 999 790	3·77	0·999 999 903	3·87	0·999 999 956	3·97	0·999 999 980
3·68	0·999 999 805	3·78	0·999 999 910	3·88	0·999 999 959	3·98	0·999 999 982
3·69	0·999 999 820	3·79	0·999 999 917	3·89	0·999 999 962	3·99	0·999 999 983

For more detailed information, see *Chambers Six Figure Mathematical Tables*, L. J. Comrie, Vol. 2, p. 518 (Chambers, 1949).

INDEX